Analysis of Variance and Functional Measurement

Analysis of Variance and Functional Measurement

A Practical Guide

DAVID J. WEISS

2006

OXFORD
UNIVERSITY PRESS

Oxford University Press, Inc., publishes works that further
Oxford University's objective of excellence
in research, scholarship, and education.

Oxford New York
Auckland Cape Town Dar es Salaam Hong Kong Karachi
Kuala Lumpur Madrid Melbourne Mexico City Nairobi
New Delhi Shanghai Taipei Toronto

With offices in
Argentina Austria Brazil Chile Czech Republic France Greece
Guatemala Hungary Italy Japan Poland Portugal Singapore
South Korea Switzerland Thailand Turkey Ukraine Vietnam

Copyright © 2006 by David J. Weiss

Published by Oxford University Press, Inc.
198 Madison Avenue, New York, New York 10016

www.oup.com

Oxford is a registered trademark of Oxford University Press

All rights reserved. No part of this publication may be reproduced,
stored in a retrieval system, or transmitted, in any form or by any means,
electronic, mechanical, photocopying, recording, or otherwise,
without the prior permission of Oxford University Press.

Library of Congress Cataloging-in-Publication Data
Weiss, David J.
Analysis of variance and functional measurement : a practical guide
by David J. Weiss.
   p. cm.
Includes bibliographical references and indexes.
ISBN-13 978-0-19-518315-3
ISBN 0-19-518315-0
1. Analysis of variance. 2. Experimental design. I. Title.
QA279.W427 2005
519.5'38—dc22   2005000606

9 8 7 6 5 4 3 2 1

Printed in the United States of America
on acid-free paper

Contents

1	Introduction	3
2	One-Way ANOVA	8
3	Using the Computer	22
4	Factorial Structure	28
5	Two-Way ANOVA	47
6	Multifactor Designs	64
7	Error-Purifying Designs	82
8	Specific Comparisons	99
9	Measurement Issues	121
10	Strength of Effect*	129
11	Nested Designs*	142
12	Missing Data*	161
13	Confounded Designs*	189
14	Introduction to Functional Measurement*	210
	Appendix A: F Table	245
	Appendix B: CALSTAT	251
	Terms from Introductory Statistics	253
	References	259
	Author Index	265
	Subject Index	267

*Chapters may be omitted or reordered without loss of continuity.
Chapter 13 has the least applicability for most researchers and is also
the most technically challenging.

Analysis of Variance and Functional Measurement

1

Introduction

ANALYSIS OF VARIANCE. The phrase sounds ominous. The word "analysis" suggests perhaps unfortunate associations with test tubes. Variance is a somewhat formal term, one whose sound is familiar from previous adventures in the world of statistics.

But whether or not previous statistical experiences were painful, analysis of variance (ANOVA) can be learned. And if one postpones (perhaps indefinitely) the proofs and algebraic derivations, it can be learned relatively painlessly. ANOVA (I pronounce this acronym with the second syllable stressed) has an odd resemblance to driving; it is easier to do than to describe, and the skill is more readily acquired through practice than through an understanding of theory.

This presentation presumes knowledge of basic statistics. A course in which elements of probability and hypothesis-testing logic were presented should suffice. If you have had that experience but memory has faded somewhat, a review of the Terms from Introductory Statistics (see p. 247) may be helpful. Terms included in that glossary appear in boldface type when they first occur in the text. The vocabulary of ANOVA will be further emphasized by the use of SMALL CAPITALS as important terms are introduced.

I employ a classical approach to hypothesis testing, in which the researcher sets a significance level for each test prior to examining the results. The American Psychological Association does not share this perspective, preferring to ask investigators to report the significance level corresponding to the obtained statistic. Either approach is compatible with the text.

You get a maximal return for learning ANOVA. It is a most powerful and versatile technique; since the late 1940s it has been the primary statistical tool of behavioral psychology. For controlled experiments and the causal inferences they allow, ANOVA remains the most natural approach. What you must learn, on the other hand, is relatively limited. The more complex analyses are simply generalizations of the simpler ones. Once the fundamental concept of partitioning **variance** is mastered, successively more sophisticated experimental designs can be analyzed.

In the everyday world of the practicing scientist, ANOVA is done on a computer. Accordingly, this text will send you to a computer soon after you have performed a few analyses by hand. But omission of the manual-labor phase will inhibit your developing the intuitions that are needed to identify erroneous results stemming from incorrect data entry or other, less common, computer problems. All of the analyses described herein can be performed with the Windows programs in the CALSTAT series accompanying the text. These programs operate in ordinary English. You need not speak a computer language to use them. I would encourage you to learn to write your own programs, but you need not do so to perform even quite complex ANOVAs.

In this text, I present more calculational detail and supporting details than some readers will want to give much attention to, although my view is that every word is a pearl. Material that can be skimmed, or even omitted, without serious loss of understanding is set in all-italic type.

The Model Underlying ANOVA

Anyone who gathers data notices variability. When the same object is repeatedly measured with a finely grained measuring instrument, as when I measure a child's height in millimeters, successive readings are rarely identical. If an examination of the series of measurements reveals no pattern underlying the differences in the observations, standard practice is to use the average of the measurements as an estimate of the value of the object. A sensible way to justify this averaging is to postulate that each observation is the sum of two components. One component is the "true" value of the object; I use the quotation marks to emphasize that this true value is unknowable and can only be estimated. The other component is a random component, which means it has a value that changes unpredictably from observation to observation. The employment of an averaging procedure is tantamount to assuming that on the average, the value of the random component is zero. The random element is presumed to be drawn, then, from a normal distribution with mean zero and variance σ_e^2. This random component is a convenient fiction created to explain the inexplicable inconsistencies in even the most careful measurements. A simple equation using **subscript notation** summarizes the assumption:

$$M_i = T + e_i \tag{1-1}$$

Equation 1-1 states that M_i, the ith measurement of the object, is the algebraic sum of T, the true value of the object, and e_i, the value of the "error" on the ith measurement. The term *error* is conventionally used for the random component.

The term is historically entrenched, though it is an unfortunate usage because it connotes a mistake rather than a normal aspect of the measurement process.

Equation 1-1 describes a situation that is too simple to be scientifically interesting. In the late eighteenth century, there arose a complication that should be dear to the hearts of all graduate students. The astronomy of that era required precise timing of the transit of a star across the meridian of the observatory. In 1795, the head of the Greenwich observatory fired an assistant because the assistant's times were about a half second too slow (that is, they were slower than the chief's). Somewhat later, the German astronomer Bessel read about the incident, and he began comparing astronomers. He found that even skilled, experienced astronomers consistently disagreed, sometimes by as much as a second. Bessel at first thought these interpersonal differences were constant, and he presented a "personal equation" that could be cast in our notation as equation 1-2:

$$M_{ij} = T + P_j + e_i \qquad (1\text{-}2)$$

Here P_j is the personal contribution of observer j. It soon became clear that there were also differences that depended on such complicating physical factors as the size of the star and its rate of movement, so a more complex equation was needed:

$$M_{ijk} = T_k + P_j + e_i \qquad (1\text{-}3)$$

Equation 1-3 has three subscripts, because it is the ith measurement of the kth object by the jth observer. The measurement now is held to depend upon the true value of the kth object (T_k), the contribution of the particular observer (P_j), and the random component (e_i).

Equation 1-3 is sufficiently complex to deal with behavioral experiments of substantive interest. Suppose, for example, one were studying how far people can throw various projectiles. There might be five different projectiles and ten throwers; fifty scores would be generated. Equation 1-3 would provide a model for the distance traversed by each projectile as thrown by each hurler.

Equation 1-3 is an abstract statement of the process underlying a set of data to be analyzed. While the statistical procedure does not make use of the model in a direct way, the model clarifies the goal of the analysis. The aim is to tie variation in the measurements to particular manipulations in the experiment. Specific terms in the model may be replaced or omitted; for example, if each thrower tossed only one projectile, there would be no way to isolate the effect on the measurements of the individual's strength. In that case, P_j would not appear in the model for the experiment. On the other hand, additional experimental complications would call for incorporating more terms into the equation. The throwers might be offered a systematically varied monetary incentive ($0 per meter, $1 per meter, $10 per meter). This experimental manipulation would require a term ($\$_l$) to specify the effect of the value of the incentive on each trial. Equation 1-4 incorporates the incentive effect:

$$M_{ijkl} = T_k + P_j + \$_l + e_i \qquad (1\text{-}4)$$

Additional complexity in the model reflects potential difficulties in interpreting the experimental results. Effects are not always simple. Suppose, for example, that people try harder for $10 per meter than for $1 per meter. Accordingly, the distances would be expected to be greater for the larger reward. But perhaps the largest projectile is so heavy that for most people it can't be thrown no matter how hard one tries; it simply drops to the ground. In that case, the expected effect of the incentive would be different for one projectile than for others. This is an example of an interaction. Interaction between two variables means that the effect of one variable depends on which value of the other variable is present. Formally, interaction terms in the model represent the effects of specific combinations of the model's components. The hypothesized interaction appears in equation 1-5:

$$M_{ijkl} = T_k + P_i + \$_l + T_k\$_l + e_i \tag{1-5}$$

Equations 1-1 through 1-5 are all instances of linear models, so named because they express the response as a linear combination of contributing elements. Other statistical procedures such as correlation and regression also employ linear models. There is a formal equivalence among the various procedures for analyzing linear models; this equivalence is conveyed by use of the term "general linear model" to refer to the family of procedures. One can, in fact, analyze the problems in this text with multiple regression (Cohen, 1968); experimental factors and their interactions are regarded as predictors whose contributions can be assessed just as one usually evaluates the impact of measured, noncontrolled variables. The ANOVA framework, though, is the natural one for working with designed experiments. Not only are the computations much simpler and easier to fathom but the elements included in the model correspond directly to those built into the experiment. With analysis of variance, one jointly plans the experiment and the analysis, which is, in my view, the path to fruitful research.

Use of the Model

A model equation is simply an algebraic representation of an experimental hypothesis. The researcher constructs the model as a guide; it points the way to the appropriate statistical tests. Each term in the equation corresponds to a particular statistical test; each term is a component in the ANOVA. The researcher does not know the correct model before the data have been analyzed. Typically, one begins by postulating a complex model, one with a term for each independent variable and with terms corresponding to all of the possible interactions among them. When the analysis reveals that some components make only negligible contributions to the variation in the scores, the corresponding terms are dropped from the model. The reduced model is offered as a descriptive statement about the experimental results.

In practice, researchers are seldom explicit about their use of these model equations. ANOVA procedures are routinized to the extent that one need not think about which components ought to be tested. Rather, the researcher identifies the proper analysis to be conducted on the basis of the experimental design. The model guides the tests, but it does so implicitly by providing a logical connection between the experimental design and the proper analysis. The linkage is implicit because it is common practice to learn the relationship between design and analysis without using model equations. We shall follow this practice since the analytical algorithms are, as a practical matter, independent of their theoretical underpinnings. A model may be used to summarize an investigation, but it is not required to carry out an appropriate data analysis. One merely tests for the factors built into the experiment along with the interactions among them. Standard significance test procedures tell us whether the factors have had their anticipated effects.

2

One-Way ANOVA

One-way ANOVA deals with the results of a straightforward experimental manipulation. There are several (two or more) groups of scores, with each **group** having been subjected to a different experimental treatment. The term one-way derives from the fact that the treatment for each group differs systematically from that for other groups in only one respect: that is, there is one **independent variable**. Within each group, the treatment should be identical for all members. Each score comes from a separate individual, or, stated otherwise, each individual contributes only one score.

The traditional name for an individual furnishing a score is SUBJECT. In recent years, the more egalitarian term PARTICIPANT has come to be favored. The modern term connotes a voluntary contribution to the research, a partnership between investigator and investigatee. VOLUNTEER is another label used for this usually anonymous member of the research team. All of these terms will be employed in the text.

The score is the quantified observation of the behavior under study. A score must be a numerical value, an amount of something. For the analysis to be sensible, the score should directly reflect the behavior in question; the greater the number, the more (or less, since it is the consistency rather than the direction of the relation that is important) of the particular behavioral tendency. An individual score must be free to take on any value in the defined range, controlled only by the experimental conditions governing that score. Linked measures (for example, sets of numbers that must sum to a particular value such as 100) do not qualify for ANOVA.

The **null hypothesis** is that the true values of the group means are equal. The simplest way to express the **alternative hypothesis** is to say that the null hypothesis is false. More definitively, at least one of the group means is different from at least one other (note the difference between the latter expression and the incorrect phrasing that the group means are all different).

A simple example would be a drug-dosage study in which the scores might be running times in a maze. We shall have three groups, with the members of each group receiving a particular dosage of a specified drug. Usually a researcher tries to have the same number of subjects in all of the groups, in order to estimate each group mean with equal precision. But things don't always work out as planned in experimental work; and in a one-way design, inequality presents no difficulties. For the sake of generality, then, our example will feature unequal group sizes.

The first group might consist of five animals, each of whom is given a 1-mg dose ten minutes before running. The second group might also have five animals, each of whom is given a 3-mg dose of the drug ten minutes before running. The seven animals in the third group might each get a 5-mg dose. The average running time for each group gives an idea of the effects of drug dosage.

If the scores in each group were completely distinct from those in the other groups, no further analysis would be necessary. More realistically, however, one would expect overlap among the scores. Some animals in the low-dosage group will inevitably run faster than some in the high-dosage group, even though the group means might suggest that, in general, higher doses lead to faster running. In order to answer the question of whether the group means are reliably different, one must carry out a statistical analysis in which the variability in the scores is taken into account.

The test, of course, is ANOVA. The variability in the scores is partitioned into two classes, systematic variance and error variance. Systematic variance is variability attributed to controlled elements in the experimental setting; in our example, the dosage of the drug was controlled. The primary systematic variance is that BETWEEN GROUPS. It measures how different each group mean is from the overall mean. If all of the group means were similar, they would as well be similar to the overall mean. Consequently the between-groups variance would be small.

ERROR variance is variation that the experiment does not aim to understand. This variation reflects idiosyncrasies participants bring with them to the experiment. People can be expected to respond differently because they have different histories and capabilities. Error variance is estimated from the average variance within groups of participants treated the same way. Since the participants within a group have been administered the same treatment, variation among their scores provides a measure of the magnitude of the idiosyncratic contribution. The error variance is, then, a composite determined from the variance within each group weighted by the number of scores per group. For this reason, error variance is also referred to as WITHIN-GROUPS variance. The variance diagram illustrates this.

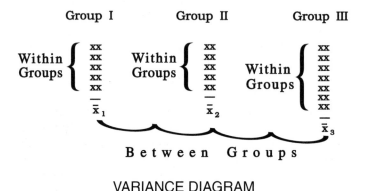

VARIANCE DIAGRAM

In terms of the model given in chapter 1 (equation 1-3), the between-groups variance includes the contributions of both the **substantive variable** (T_k) and the random error component (e_i). The within-groups term, on the other hand, contains only the error component. In this experimental design, the personal contribution of each participant (if a rat may be said to make a personal contribution) is **confounded** with the error. Because the individual makes only one response, it is not possible to identify the separate contributions of the personal component and the error component. So the error component in this design includes individual differences; both the between-groups variance and the within-groups variance include this masked contribution.

To the extent that the substantive variable, in this example the drug dosage, has a big effect, then the between-groups variance will be large relative to the within-groups variance. In contrast, suppose that the substantive variable had no effect, in other words, that the running times were not differentially affected by how much of the drug the animal received. In that case, the between-groups and within-groups variances would both simply be measuring unsystematic variation. One would expect the two variances to be comparable in magnitude. They will not be identical, of course, because different scores are being processed in the computations, and thus different instances of the random component are involved.

There are two plausible ways to compare quantities. One can examine the difference between them, which ought to be close to zero if the quantities are the same except for random error. Alternatively, one can examine the ratio, which ought to be close to one if the quantities are the same except for random error. Which method is preferable? The ratio conveys more information. An analogy may clarify this argument. Suppose I have been trying the new Pasadena diet, and I proudly report that I lost 10 pounds. Is that a sufficiently impressive reduction for you to consider recommending the diet to a friend who wants to lose weight? In order to make that judgment, you might want to know the weight I started from. If I originally weighed 360 pounds, the reduction would hardly be noticeable, but if I had originally weighed 180 pounds, the difference in my weight would be more impressive. The diet's effectiveness can be conveyed compactly by reporting the percentage of my original weight that I lost rather than the amount. With the percentage information, you don't need to know how much I originally weighed in order to evaluate the diet. Comparing two numbers via a ratio is akin to using a percentage. The ratio expresses the magnitude of the one number (the numerator) in units that are the magnitude of other number (the denominator) so that the actual values of the numbers being compared are not required to appreciate their relationship.

Following this reasoning, statisticians routinely express comparisons as ratios in the procedures they develop. The F test, named in honor of R. A. Fisher, the British agronomist and statistician who pioneered ANOVA, examines the ratio of the between-groups variance to the within-groups variance. If the between-groups variance is large compared to the within-groups variance, then this ratio, called the F ratio, will be large; this would be evidence supporting the idea that drug dosage makes a difference. On the other hand, if the treatment had no effect, then the F ratio would be expected to be approximately one; that is, the between-groups variance should be of about the same size as the within-groups variance.

Chance fluctuations that in terms of the model (equation 1-3) are randomly varying values of the error component e_i may affect either of the two critical variances and thus affect the F ratio. Therefore, the **probability distribution** of the F ratio, under the assumption that the two involved variances are in truth of the same magnitude, has been worked out. This distribution furnishes the entries in the F table. The tabled value is employed as a **critical value**, or criterion. If the F ratio obtained from the data is larger than the tabled value, then the between-groups variance is deemed large relative to the yardstick of the within-groups variance. In other words, a large, or significant, F ratio is evidence that the group means are not all the same. An obtained F ratio not exceeding the critical value suggests that the group means are not reliably different. Alternatively and equivalently, if the **p value** associated with the obtained F ratio is less than the designated significance level, the difference between the group means is deemed significant. Because the fundamental quantities leading to the ratio are variances, all of which must be positive, directions of differences between means are not preserved. Therefore, all tests employ only the upper tail of the F distribution. F tests are treated as **one-tailed** even though the alternative hypothesis is nondirectional.

Randomization and Independence

Suppose a woman receives the dreaded news that she has breast cancer and asks you for advice about where to seek treatment. One element in the response might be an evaluation of the survival duration for patients who have gone to various hospitals. If this information were available, significant differences might well have life-or-death implications. It would seem natural to avoid a facility whose patients did not live a long time after treatment. Unfortunately, this natural conclusion might be the wrong one, and your advice might well prove fatal. You have made the assumption that all patients are equivalent. Suppose, for example, the suspect hospital was known among local physicians as the best, and accordingly physicians directed their most seriously ill patients there. The statistical evaluation is useless because we don't know whether the patients in the different institutions are comparable. Experimental control, as this realistic example vividly demonstrates, is no mere technical nicety.

An experimental comparison depends upon the idea that consistent differences between scores from participants in various groups are not the result of preexisting differences. ANOVA can tell us whether group means are reliably different but not whether the differences were caused by the experimental treatment. In order to make the inference that group differences are linked to treatment effects, the researcher must see that prior to treatment the groups are comparable in whatever respects are crucial. The easiest and best way to achieve this goal is to randomly assign participants to groups. Every subject should have the same probability of being assigned to any of the experimental groups. While randomization cannot guarantee that the groups are indeed equivalent prior to treatment, it does insure against bias, that is, stacking the deck in a particular direction.

Sometimes practical constraints prohibit a random assignment. This situation occurs when the variable is classificatory rather than experimental. For example, if gender or age is the variable of interest, the assignment of participant to group can hardly be determined randomly. In such a situation, the problem is that one cannot say with confidence that observed between-group differences are related to the variable of interest. There may be a hidden variable, such as weight or height or years of education, that is truly responsible for the experimental effect. The classificatory variable, sometimes called a *subject variable*, may be naturally confounded with a hidden variable that, although logically distinct, is associated with the classification. Random assignment minimizes the chance that such a concomitant variable will confuse the researcher.

It is worth noting that this problem is not related to ANOVA but to the design of the experiment. The statistical analysis is neutral. It is designed to tell you whether the average scores in the experimental groups are reliably different. The issue of what the scores mean or of whether the numbers are meaningful at all is not in the domain of statistics but of experimental logic.

Experimental logic also demands that the observations be independent of one another. This means primarily that the researcher must collect each score in the same way. In a practical sense, of course, it is not possible for an experimenter to be equally tired or for the apparatus to be in the same condition for all observations. Once again, randomization comes to the rescue. By interweaving the subjects from the various groups according to a random sequence, the researcher avoids biasing the results.

How does one achieve randomization? Suppose it is desired to assign five subjects to each of four experimental groups. As volunteers report in from the introductory class pool, each one is placed in a particular group according to a predetermined scheme. Since there are four groups, regard the subjects as coming in sets of four. For each set, shuffle the four names, or more conveniently the index numbers 1, 2, 3, and 4, in a hat (hats are traditional for shuffling, though no one I know owns a hat these days, so you may have to improvise). The first one drawn goes into group 1, the second into group 2, and so on. Repeat this shuffling process five times, in each case using a separate shuffle. Alternatively, use a computer program to generate random permutations to accomplish the shuffling. A major advantage of the permutation scheme, as opposed to independent randomization as each subject comes along, is that equal group sizes are automatically attained as each permutation is implemented.

The F Table

The F distribution is actually a family of distributions. There are two parameters that serve to distinguish the members of the family. Each F distribution is characterized by two **degrees of freedom** (df) values. The phrase "degrees of freedom" has little explanatory or mnemonic value, but it is unfortunately embedded in the literature. The DEGREES OF FREEDOM FOR NUMERATOR heading and the DEGREES OF FREEDOM FOR DENOMINATOR headings guide the table user to the proper critical values, as do the coordinates on a map. The values of these parameters are determined by the structure of the experiment. The degrees of freedom for numerator are one less than the number of groups (for our drug experiment, $df_{num} = 2$). The degrees of freedom for denominator are computed by subtracting the number of groups from the total number of scores (for our drug example, $df_{denom} = 14$).

A researcher arbitrarily chooses a **significance level**, or in other words, determines the probability that the true group means will be declared different when they are in fact the same. This misjudgment is called a **Type I error**. This significance level is usually set conventionally at either .05 or .01, though there is no logical necessity governing the choice. Throughout this text, the .05 level is presumed to have been selected. The significance level also determines the critical value of F. Examination of the F table appendix A reveals that the more strict the criterion (that is, the lower the significance level chosen), the larger the obtained F ratio must be in order to exceed the tabled critical value. This means that the choice of significance level plays a role in determining how large an observed difference among group means will be required before those group means are pronounced different. For the drug experiment with $df = 2, 14$, the table shows the critical value for F for the .05 level of significance to be 3.74, while that for the .01 level is 6.51.

Power

The significance level also affects the probability of a **Type II error**, that is, failing to confirm a true difference among the group means. The capability of detecting a difference is known as the **power** of the statistical test, and obviously it is desirable for a test to be powerful. The less stringent the significance level (that is, the larger the value of α), the more powerful the test is because a smaller F ratio is required in order to attain significance. But manipulating the significance level to gain power is a dangerous game because there may in fact be no true difference among means, and a higher significance level increases the risk of a Type I error.

Fortunately, power may also be increased by means that do not affect the Type I error rate. The most fruitful ways under the control of the researcher involve reducing the within-groups variability, thus producing a smaller denominator for the F ratio. Possibilities include choosing participants to be homogeneous, specifying experimental instructions and procedures carefully so that all observations are generated under the same conditions, and eliminating or controlling (via the more complex experimental designs to be encountered in later chapters) extraneous variables such as time of day or temperature of the room.

Power can also be increased by increasing the number of participants because the corresponding increase in the denominator df means that a smaller F ratio is needed to reach statistical significance. However, in addition to the economic burden, there is another drawback to solving the power problem by running hordes of subjects. This drawback can be illuminated by considering the extreme case. Suppose there is a very small but true difference among the group means. The greater the df for the denominator, the more likely this difference is to be significant. With a very large number of subjects, even a tiny difference may prove significant. But few researchers want to discover or confirm experimental manipulations that produce small differences. Rather they want to demonstrate potent treatment effects. Because ANOVA addresses the question of reliability rather than magnitude of effect, a significant F ratio does not necessarily imply an impressive result. Most researchers are wary of studies that employ huge numbers of participants to obtain small but significant F ratios for fear that the effects are small and therefore perhaps not worthy of attention. So an experimenter should run enough, but not too many, subjects. This can be a tricky problem. More will be said on this matter in chapter 10, where we discuss strength of effect in detail.

Computation

Although it is feasible to determine the between-groups and within-groups variances with formulas based on the usual definition of a variance (the only complication is that when group sizes are unequal, each group's contribution must be weighted proportionately to its size), a much more convenient computational scheme is available. With this algorithm comes a standard format for presenting the results and some new terminology.

ANOVAs are customarily presented in a table, and the table has conventional headings for its columns. The term SOURCE is used to refer to the particular source of variation for each row in the table; at this point, our sources will be between groups and within groups. Later, as we encounter more complex designs in which variability is further partitioned, there will be more sources and thus more rows. Each source's *df* are given, and then we come to the crucial sum of squares (SS) and mean square (MS) columns. The mean square for a source is the (weighted) average variance for that source, and it is the mean squares that are used to construct the *F* ratio that appears in the final column. The sum of squares for each source is the quantity computed directly from the data. From each SS the corresponding MS is derived. Sums of squares merit an entry in the table, though, because they are not merely an intermediate calculation on the way to the mean squares. The theoretical importance of the SSs is that they are additive. This additive property means that sums of squares for the various sources add up to the total sum of squares for the whole experiment. Similarly, when we do the further partitioning called for by more complex designs, it is the SS that is partitioned. Mean squares, on the other hand, are not additive.

An Example of One-Way ANOVA

The scores in the table represent running times from three groups of animals in a drug-dosage study. Placing the data in tabular format is a worthwhile step in the analytic procedure. Neatness doesn't exactly count, but it is useful to avoid getting lost in a maze of numbers.

Running Time Scores, in Seconds

Group 1	Group 2	Group 3	
18	16	10	
23	16	17	
14	11	8	
16	18	12	
23	14	14	
		7	
		11	
$t_1 = 94$	$t_2 = 75$	$t_3 = 79$	$T = 248$

There are three numbers to calculate; I refer to them imaginatively as (1), (2), and (3).

(1) ΣX^2: Each of the scores is squared, and these squares are then summed. This is usually the most tedious step in any ANOVA, although it is not too unpleasant if your calculator has an "M+" key. Simply punch the "×", "=", and "M+" keys in sequence after you enter each of the scores, and the ΣX^2 should appear when you press the memory recall key. This convenience does not add much to the price of a calculator, and it should swing your buying decision.

(2) T^2/N: T is the grand total, the sum of all of the scores. (T should be called ΣX for consistency's sake, but the label is traditional.) N is the number of scores that went into T, that is, the total number of scores in the experiment.
(3) $\Sigma(t_j^2/n_j)$: t_j is the total for the jth group, and n_j is the number of scores in that jth group. Compute each t_j by summing the scores in each group separately. Each group total is squared and then divided by the number of numbers that contributed to the total. The results of these divisions are then summed.

The defined quantities are calculated from the data:

(1) ΣX^2: $18^2 + 23^2 + 14^2 + \cdots + 14^2 + 7^2 + 11^2 = 3{,}950$
(2) T^2/N: $248^2/17 = 3{,}617.88$
(3) $\Sigma(t_j^2/n_j)$: $(94^2/5) + (75^2/5) + (79^2/7) = 3{,}783.77$

Next, the calculated quantities are used to generate the numbers in the SS column of the ANOVA table. As SSs are literally *sums* of *squares*, they are necessarily positive. It is inevitable that (1) should be the largest calculated quantity and (2) should be the smallest. If an arithmetic error causes a violation of this ordering, a negative SS will result. The good news is that at least that error will be spotted (for me, it's quantity [1] on which my calculator is most likely to fail).

ANOVA Table

Source	df	SS	MS	F
Between groups	2	(3) – (2) = 165.89	$\dfrac{SS_{bg}}{df_{bg}} = \dfrac{165.89}{2} = 82.95$	$\dfrac{MS_{bg}}{MS_{wg}} = \dfrac{82.95}{11.87} = 6.99*$
Within groups	14	(1) – (3) = 166.23	$\dfrac{SS_{wg}}{df_{wg}} = \dfrac{166.23}{14} = 11.87$	

The asterisk sitting proudly beside the F ratio denotes significance at the researcher's chosen level. In an actual table, of course, only the numerical values appear, not the formulas or intermediate calculations.

Numerical Details

Numerical accuracy is certainly a goal worth striving for. In presenting results for public consumption, though, one cannot expect ten decimal places to be tolerated. One must round the numbers to be presented in the ANOVA table. It is customary to report sums of squares and mean squares to one or two decimal places and F values to two places. Maximal accuracy is achieved by maintaining as many decimal places as your calculator will hold until the computations are complete; only then should rounding be done. Consequently, the reported F ratio occasionally will have a slightly surrealistic quality in that the ratio of the reported mean squares does not precisely yield the F given in the table. It is more important to provide a correct statistic than to appear consistent. One should avoid rounding at intermediate stages.

The Responsiveness of ANOVA

A further example will serve to clarify the way the statistical test is sensitive to the data. The table shows three groups of numbers drawn from a random number table.

Scores from Random Number Table

Group 1	Group 2	Group 3
4	13	14
11	7	2
13	6	5
16	16	4
1	4	6
2	12	10

The ANOVA computations yield an F of 0.46. Since F is less than one, no table is necessary to verify the nonsignificance of the between-groups effect; in fact, F ratios of less than one are customarily reported simply as "$F < 1$." This result is hardly surprising, considering the origin of the scores.

Now modify the scores by adding 10 to each score in group 1 and 5 to each score in group 2. Recompute the ANOVA. This time the F ratio leaps to a value of 7.13*. This statistically significant F ratio reflects your having imposed a between-groups effect onto the data. Notice that MS_{wg} (27.0) for the modified scores is the same as for the original scores; this reflects the fact that the variance of the scores within each group did not change.

Next, return to the original scores and apply a different modification by adding 10 to the last score in each group. Once again, recompute the ANOVA. Now the F ratio (0.29) is even smaller than that for the original scores. The reason is that this second modification has increased the within-groups variance, but it has not increased the between-groups variance since the group totals are as different from one another as they were before.

A Test of the Grand Mean

Summing the degrees of freedom in the ANOVA table yields $N - 1$, one less than the number of scores. Since the rule for generating degrees of freedom is that each score produces one degree, there must be an element missing from the table. The missing link is a seldom-tested source that compares the overall mean response to zero. The sum of squares for this source is T^2/N, which we know as (2), and since SS_{mean} has only one degree of freedom, the corresponding mean square is also given by T^2/N. This mean square may be compared to the mean square within, with an F ratio being formed and tested for significance in the standard way. If the F ratio proves significant, it is interpreted as evidence against the null hypothesis that the grand mean of the scores is zero.

It should be clear why this source is usually omitted from the ANOVA table. The value of the average response is rarely of interest to the researcher; what is of concern is whether the various treatments have had differential effects. The only practical situation in which the test for the mean is likely to be useful is when the data are comprised of difference or change scores. Consider a project in which two different programs for weight loss are evaluated. Primary interest would surely be in whether one program produces more weight loss than the other; with individual losses as the scores, the ordinary between-groups F ratio is addressed to that question. But also of interest might be the question of whether the programs are effective at all. If the average weight loss is not reliably different from zero, then the reduction programs must be regarded as failures. The appropriate test of this question employs the F for the mean.

One may also test the null hypothesis that the grand mean is equal to some other predetermined constant, K, rather than zero. In this case, the numerator of the F ratio is modified to incorporate the constant:

$$SS_{mean} = N \cdot (T/N - K)^2$$

Notice that T/N is simply the grand mean, and so if $K = 0$ the expression reduces to T^2/N. I have never seen this null hypothesis tested in print (please don't deluge me with citations), so the derivation is probably not of great importance. Still, it's nice to know where the missing *df* goes and to appreciate its meaning.

Exercises

You will see that I use varied formats for presentation of the data. This is a deliberate maneuver to prepare you for the wonderful variety used by researchers as they scribble their scores. However, your ANOVA tables should rigidly follow the format given in the text; it is considered anarchy to try a different format for the table. Use the .05 level of significance as a default throughout the text.

2-1. I conducted an experiment on two sections of my introductory statistics class. One section (with four students) had a graduate assistant, while the other (with eight students) did not. Determine whether the assistant makes a difference. The scores are from each student's final exam.

Section 1 (with assistant): 70, 50, 60, 60

Section 2 (without assistant): 30, 20, 40, 10, 50, 30, 20, 40

2-2. The following scores represent the number of errors made by each person on a verbal learning task. Each person was assigned to one of four study groups. Test the hypothesis that the different study groups all produced the same average number of errors.

Group	Error scores
1	16, 7, 19, 24, 31
2	24, 6, 15, 25, 32, 24, 29
3	16, 15, 18, 19, 6, 13, 18
4	25, 19, 16, 17, 42, 45

2-3. Students taking Psych 205 were randomly assigned to one of three instructional conditions. The same test was given to all of the students at the end of the quarter. Test the hypothesis that there were no differences in test scores between groups.

Group	Test Scores
Lecture	10, 13, 3, 38, 11, 23, 36, 3, 61, 21, 5
Programmed instruction	8, 36, 61, 23, 36, 48, 51, 36, 48, 36
Television	36, 48, 23, 48, 61, 61, 23, 36, 61

2-4. A professor of psychiatry evaluated three of her trainee therapists by asking their patients for self-reports of their perceived growth (0 = no growth; 20 = maximal growth) during the course of therapy. Test the null hypothesis that the trainees were equally effective. Also evaluate the professor by testing the null hypothesis that on the whole the patients experienced no growth at all.

Scores from Patients of Trainees

Trainee A	Trainee B	Trainee C
2	0	3
5	2	4
0	1	2
1	0	1
3	1	

2-5. Bozo, a statistically minded clown, decided to evaluate five new routines he had created. One of his fellow clowns, Dumbo, contended that children laugh at anything done by a clown, but Bozo argued that some ideas are more hilarious than others. Bozo went to the Stoneface Elementary School and successively gathered 5 groups of 4 children. For each group, Bozo performed 3 minutes' worth of one of the routines while Dumbo counted the number of laughs emitted by each child. Whose point of view, Bozo's or Dumbo's, do these comical results support? Each score given is the number of laughs by one child in response to the routine.

ANALYSIS OF VARIANCE AND FUNCTIONAL MEASUREMENT

Number of Laughs Emitted by Groups of Children in Response to Comedy Routines

Group	Pies in faces	Monkey imitation	Pratfalls	Snappy insults	Revolting smells
Group 1	12	7	20	4	15
Group 2	13	14	25	2	18
Group 3	11	22	17	0	23
Group 4	25	8	22	8	17

2-6. The Committee Against Violent Television charged that Network 2 was the most violent of them all; Network 2 responded that they were no worse than their competitors. The committee assigned watchers to count the number of brutalities per evening for 5 days. Evaluate the data to determine if the networks are equally culpable.

Number of Brutalities Counted on Television Networks per Evening

Network 1	Network 2	Network 3
18	42	32
32	73	28
23	68	17
16	32	43
19	47	37

Answers to Exercises

2-1.
Source	df	SS	MS	F
Between groups	1	2400.00	2400.00	17.14*
Within groups	10	1400.00	140.00	

2-2.
Source	df	SS	MS	F
Between groups	3	513.97	171.32	2.06
Within groups	21	1749.29	83.30	

2-3.
Source	df	SS	MS	F
Between groups	2	3141.93	1570.96	5.85*
Within groups	27	7249.53	268.50	

2-4.

Source	df	SS	MS	F
Between groups	2	7.76	3.88	1.89
Mean	1	44.64	44.64	21.73*
Within groups	11	22.60	2.05	

2-5.

Source	df	SS	MS	F
Between groups	4	721.30	180.33	7.21*
Within groups	15	375.25	25.02	

2-6.

Source	df	SS	MS	F
Between groups	2	2476.13	1238.07	8.41*
Within groups	12	1767.60	147.30	

3

Using the Computer

The computer is dangerously seductive for a student. The temptation is to use the machine to do the computational work, saving the conceptual components of an analysis for the human member of the team. That's certainly not a poor general strategy, and researchers routinely follow it. I believe, though, that the manual part of the learning process is crucial, and that if one skips it, understanding will always remain at a superficial level. The machine will take on magical properties, and an analysis for which there seems to be no appropriate program will be an analysis that does not get done. Even worse, an experiment's feasibility may be determined by the availability of a program that can analyze the data. Just as it was good for Mr. Lincoln and me to walk twenty miles in the snow to get to school, it is good for a student to do problems by hand first.

Pragmatically, too, it makes sense to learn an analysis before you try it on a computer. The principal stumbling block in using a program is the language of the interface. The user needs to know the terminology the program employs to request information, and also needs to be able to interpret the output.

Modern programs are better than ever. A well-written Windows program, such as those in the CALSTAT series, requires no manual and has almost no learning curve. One must know the fundamental operations common to all such programs—selecting with the mouse, opening a file, and so on—but specifics for a given analysis should be obvious to a user who knows the statistic being computed. Fully indexed online help, with searching via keywords, is a standard feature available by clicking from a menu. Context-sensitive help is often available as well; pressing the F1 key will elicit information relevant to the current operation.

There are two basic types of programs available for analysis of variance. Large commercial packages, such as SPSS or SAS, can handle many statistical procedures, of which ANOVA routines form a subset. In general, these programs are (a) very powerful, in that they can cope with a wide variety of designs; (b) accurate, because (at least for the most commonly used procedures) they have been checked out by hosts of previous users; (c) expensive, because of the generality; and (d) relatively difficult to use, because they must be sufficiently general to handle so many disparate analyses and perhaps because they are written for sophisticated users. If you have access to such a package and are comfortable with its operation, then go ahead and use it. Experience gained with CALSTAT transfers nicely, although you may be spoiled by CALSTAT's friendliness and resist the switch. Quite a few of the analyses offered by the various CALSTAT programs are not available in the commercial packages (and vice versa, of course).

The alternative is the smaller, specialized program. The CALSTAT series exemplifies this approach. User friendliness has been a paramount consideration. This term means that the program helps the user enter data correctly and tries to prevent errors. Those errors that cannot be prevented, such as entering a value incorrectly or omitting a score, are easily corrected. All of the programs use a similar interface. The user specifies the design structure by responding to questions. Data are entered via a labeled, scrollable grid. Corrections in data entry require only a few clicks (check out the Data Editing menu). When it has been determined that the data were entered correctly, another button click either allows setup of the design or immediately launches into calculations and produces the appropriate table. Graphing of means is available. Results may be examined on the screen, printed directly, or put into a Word document. The data can be placed into a file (perhaps for reanalysis at another time or with a different procedure) at any time from the grid.

> *If the idea of a computer file is unfamiliar, the saving options offered by the program will seem mysterious, so an overview may be helpful. "File" is a conventional term for a stored set of information, analogous to a file in a cabinet. A file is created within a particular program and may be, or may not be, usable by another program. Some programs, such as Microsoft Word, use a proprietary format for their files. The CALSTAT programs communicate directly with recent versions of Word, producing documents with the standard .doc extension. If the communication does not work, you can use the Copy command available through the Edit menu to place a single piece of information on the Clipboard, from which it can be subsequently pasted into Word. Results generated by the various output options can be embedded into a file such as myfile.doc; that file can then be modified using Word's full word-processing capabilities. CALSTAT programs save data in .dat files, such as mydata.dat. Since these files contain only numbers, they can be read by other programs, such as Microsoft Excel. File names should be unique names, with an extension indicating the type of information in the file. The extension will usually be suggested by the program; subsequent usage will be simplified if you adopt the suggestion.*

Files may be placed in folders or subfolders, which in turn are stored on a particular drive. The hard drive in an individual computer is usually configured with the name C:. Subfolders are always almost used on the hard drive, so a data file might be entered as C:/CHAPTER6/6-2.dat. The slashes separate drive from folder and folder from file. The floppy drive (the one that accepts the plastic diskette) is usually called A:. An output file prepared by Word might be A:/6-6.doc. It is a really, really good idea to employ file names that have mnemonic value.

Installation Instructions for CALSTAT

The CALSTAT package installs through a standard Windows setup program. Close other applications, then insert the CD-ROM into the appropriate drive (we'll call it E:). From the Start Menu, click on Run . . . Type in E:/Setup (use the correct drive name for your system). Alternatively, you may double-click on Setup.exe (*not* Setup.Lst) from Windows Explorer (having opened drive E:).

By default, Setup will offer to install the programs in the C:/Program Files/ CALSTAT directory; you may choose another location, but there probably is no reason to do so. Icons will be automatically created in the CALSTAT group.

Requirements and Properties of the CALSTAT Programs

Installation of the programs requires Microsoft Windows 95 or later. To use the Word options for printing, Word97 or later is needed (printing may be done without Word). ONEWAY and COMPARISONS allow any number of groups, FACTORIAL ANOVA allows up to eight factors, FUNCTIONAL MEASUREMENT allows up to six factors, SNAPSHOT allows up to three factors; all of these programs are limited to 16,352 scores per analysis.

Exploring the Programs

The first time you use a new program, it is worthwhile investing a few minutes to explore its capabilities. Read the Purpose statement offered through the Help menu on the opening window. A description of what the program does is given, along with information about terminology and limitations. Check out the Set Preferences menu on the opening window as well; these are toggles allowing you to configure the program's appearance and operating modes. Toggling means that the option continues to be on until you toggle it off. The program will remember your settings even after you have turned off the computer. There may be specialized options that you are not ready to use until you've read more of the text. Although most requests made by the program are self-explanatory, you may occasionally want to explore the Contents of the Help system for further information. This information is offered from every screen.

Below is the opening window for the ONEWAY program. I clicked on the Set Preferences menu item to expose the options this program offers at the outset. If I were to toggle Run Maximized by clicking on those words, a check would appear beside that option. Thereafter, all windows within the program will appear in full-screen mode until the check is removed. Full-screen mode is nice if you like to see things Really Big. If you instead enlarge the view by clicking the maximizing button at the upper right of a window, only the current window is shown in full-screen mode. Clicking Check Keyboard Input for Outliers invokes a routine that will scan the data after you enter all of the scores; if a value is far removed from the others, the program will ask you to examine the aberrant value. That option helps to catch data entry errors, and as a certified poor typist, I always invoke it. I turn on the Include Test of Mean Response option only when I plan to incorporate that extra line in the ANOVA table (say, for exercise 2-4).

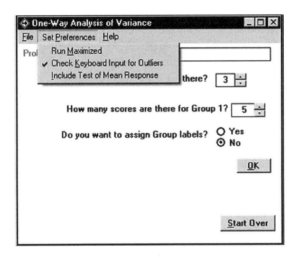

After you answer the questions about the structure of the design, the program will tell you how many scores it is expecting and ask you to confirm that the number is what it ought to be. The Data Entry window will appear next. Here, I have begun typing in scores for exercise 2-1. After typing a score, I hit the Enter key, and the focus moves down to the next row. The labels "Assistant" and "No assistant" are there to help guide the entry process; those labels appear because I answered "Yes" to the Group labels question on the previous window. If I had instead answered "No," then the program would have used a default label indicating Group 1, Group 2, and so on. After data entry is complete, clicking on Data OK will check that a score has been entered in every row (and perform the check for outliers if that option has been chosen). Then the buttons that are grayed out in the window depicted below will become available.

26 ANALYSIS OF VARIANCE AND FUNCTIONAL MEASUREMENT

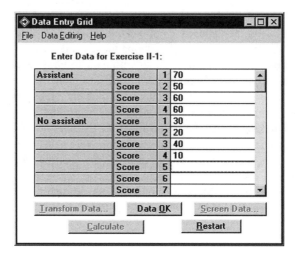

Let's examine the output from the ONEWAY program, here using the data from the demonstration problem in chapter 2.

The program displays a conventional F table, but does not append an asterisk to the significant F ratio. Instead, a p value is provided that allows you to determine significance at your chosen level. You can perform several follow-up analyses that we haven't discussed yet by clicking on their buttons, and can also look at a graph or at various descriptive statistics for the individual groups.

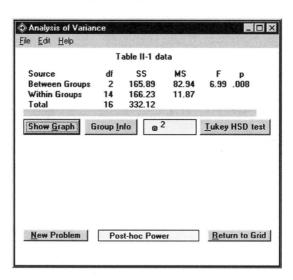

Because the CALSTAT programs are so intuitive, this can be a very brief chapter. However, if you are new to computers or to Windows, you may want to consult one of the many excellent books available. An alternative is to have an experienced colleague spend a short while by your side at a machine; it really doesn't take very long to learn how to run the programs. You need not learn "to program"; you are learning to be a user (this is not a pejorative term).

Exercises

3-1. Get started by using the computer to analyze the exercises from chapter 2. You will, not surprisingly, be using the program called ONEWAY. The easiest way to launch a program is by double-clicking its icon. The icon for ONEWAY looks like—well, you'll find it. After the program loads, it's a good idea to familiarize yourself with its options. Pull down each menu to see what's available on it. Departure from a program is via the File menu; it is a Windows convention to place an Exit option there.

3-2. Use the computer program RANDOM PERMUTATION to develop an assignment scheme for an experiment planned to have ten subjects in each of six conditions. Would there be a problem if only thirty-six subjects proved to be available?

Answer: I can't show "the" answer since the permutations are constructed randomly; here is one possibility. According to my random result, the fifteenth subject will be assigned to the third condition.

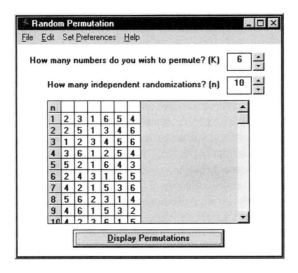

You should have ten sets of random arrangements of the numbers 1 through 6. Each set dictates how the next six subjects will be assigned to the experimental groupings. In that way, should something happen that causes the experiment to terminate before its planned completion (for example, if the researcher runs out of money or time), the subjects already run will have been properly assigned to the six conditions. So if only 36 subjects were available, you would simply run the first six sets of random arrangements.

4

Factorial Structure

One-way ANOVA is a fine little technique, to be sure, but it's really just a generalization of the t test to more than two groups. What makes the analysis of variance framework so powerful is the way it extends to combinations of experimental treatments. We now consider experiments in which there are multiple independent variables. The additional information generated by combined treatments allows the examination of forces acting in concert. In the real world, behavior is influenced by many stimuli that operate simultaneously. In the laboratory, the researcher attempts to eliminate the effect of almost all of these possible forces, and at the same time to control the presentation of a small number of designated stimuli. The FACTORIAL DESIGN is the researcher's plan for combining stimulus elements.

The simplest factorial design consists of two factors. A FACTOR is an experimental treatment under control of the researcher. The terms VARIABLE and WAY are used synonymously with factor; thus, we may speak of one-way ANOVA or two-factor ANOVA. Consider an experiment where the focus is on which toys are selected by children in a play setting; the outcome measure, or **dependent variable**, might be the amount of time the subject plays with the toy. The toys are geometric objects: cubes and balls. The toys differ in size, color, and material as well as in shape. Suppose the researcher's interest is in how shape and color affect playing time. These two aspects are designated as factors. The next decision concerns the choice of LEVELS for each factor. The levels are the different values chosen by the researcher to constitute the factor. For the shape factor, the levels might be cube and ball. For the color factor, the levels might be red, blue, and yellow. This selection would give rise to a 2 × 3 design, yielding six stimulus combinations of interest. The shape factor has two levels, the color factor three.

Although the stimuli are constructed in terms of the two aspects, each one is an integral unit. The set consists of a red cube, a red ball, a blue cube, a blue ball, a yellow cube, and a yellow ball. Schematically the combinations can be seen as:

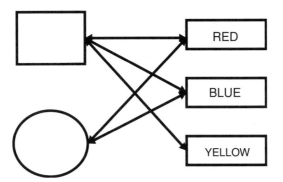

The arrows in the picture illustrate of the reciprocal nature of the relationship among factors. Although the verbal description suggests the colors modify the shapes, the relationship is absolutely symmetrical; we could as well say that the shapes modify the colors. The factors are said to be CROSSED; the choice of which to call the first factor and which the second is arbitrary. Every level of one factor is paired with every level of the other factor. Another way to represent the crossing relationship in visual terms is with a two-way table:

	Cube	Ball
Red	11	21
Blue	12	22
Yellow	13	23

Here we introduce subscript notation, a convenient way of denoting a particular CELL, or treatment combination, within the factorial design. The code should be understood as follows: 11 = first level of first factor, first level of second factor; 12 = first level of first factor, second level of second factor; ... ; 23 = second level of first factor, third level of second factor. The term subscript is used because the cell is often symbolized as X_{ij}. Just as the choice of which factor is first is arbitrary, so is the choice of which level within a factor is designated as first or second. Once the subscripts are put into place, though, the positions of factors and levels are fixed. In the table above, which factor is Factor One?

What of the other aspects in which the toys vary, aspects not under consideration in the study? The toys inevitably have other properties whose impact is not of current interest. One possibility is to allow the toys to vary haphazardly on the other dimensions. Some of the toys might be big, some small; some might be made of wood, some of Nerf material (whatever that is). If these extraneous properties do in fact affect the child's selection, their contribution to the data will be in the form of noise. There will be uncontrolled variability, which inflates σ_e^2. Another disturbing possibility is that the "irrelevant" aspects of the toys might vary systematically. For example, the Nerf balls might all be yellow, while the wooden balls might all be red. This kind of defective design is referred to as CONFOUNDED, because it makes it impossible to distinguish the effect of one variable from that of another. Does the child like the feel of wood or the color red? When the two aspects covary, one cannot tell. While a complete confounding is an unlikely error for a researcher to make, partial confounding is easy enough to slip into accidentally. Most of the Nerf balls might just happen to be yellow. No one planned it that way; that's just how the toys happened to come from the store.

A more effective approach is to hold the properties not included in the factorial design constant. Thus, all of the balls might be wooden. This plan is optimal, but sometimes it is not feasible to control everything about the stimuli. Let's change the study so that the stimuli to be evaluated are not toys, but other children. As in the previous experiment, we want to measure how much time the child plays with the stimulus. The other children may vary in size, color (in this context, we would use the term *ethnicity*), and shape just as the toys did. Our focus might be on the factors size and ethnicity. The two studies are quite similar, yet there is a fundamental difference about the stimuli. When people are the stimulus objects, how do we control the extraneous ways in which they vary? The ways are almost too numerous to catalog, and even if we could list the attributes, how would we hold the chosen values constant? Here the researcher's most direct strategy is to control the obvious differences, such as age and sex (e.g., the possible playmates are five-year-old boys who vary in size and ethnicity), and to hope the uncontrolled dimensions do not inflate the variability too much. The alternative is to incorporate additional factors into the stimulus design, thus increasing the complexity of the experiment (and the analysis, but we'll be ready for that before long); still, it is not always easy to identify or control the contributing elements. In this experiment, how would you deal with personality variables?

Interaction

The pairing of factors allows the researcher to broaden the conclusions about how a single factor operates. In the toys example, a one-way experiment in which shape was varied but color was fixed would allow a conclusion about shape limited to the single color. By incorporating the additional variable, the conclusion about shape is more interesting because the impact is more general. It is more useful to know that children prefer cubes to balls than to know that they prefer red cubes to red balls.

Sometimes, though, the results are not so easy to express. What if children prefer the red cube to the red ball, but spend more time with the yellow ball than the yellow cube? That is, the impact of one variable is not consistent across the levels of the other. This kind of outcome is known as an INTERACTION.

When two experimental treatments have been simultaneously applied, there are three questions of interest. Naturally, we are interested in the efficacy of each treatment. But what about their joint effect? Consider a chemical analogy. The element sodium is impalatable, and the element chlorine even more so, but a chemical combination of the two, sodium chloride, is table salt. The impact of the combination is not predictable on the basis of knowledge of the separate effects of the components. A measure of interaction is the unique contribution of factorial ANOVA; information is extracted that would not be available if the two treatments had been administered sequentially.

A colleague is interested in the effects of recreational drugs on driving performance. In the laboratory, he administers doses of alcohol and marijuana and assesses their effects by looking at reaction times (RT) to a simulated stoplight. Most people would guess that each drug would slow down the reaction, but what happens when a person gets both drugs? (This is a practical problem in some communities.) Suppose that a person who has received a dose of alcohol is on the average 50 milliseconds slower than a sober subject, while a person who has received a dose of marijuana is on the average 30 milliseconds slower than his straight counterpart. Will the individual effects simply add? ADDITIVITY of effects is the antithesis of interaction. Additivity would imply that a volunteer who has received both drugs is on the average 80 milliseconds slower than the volunteer who has received a placebo, or fake drug. If that occurred, we would say that the effects simply add; the joint effect is predictable from knowledge of the effects of the components.

Interaction, or nonadditivity, means that the scores for doubly drugged subjects are different from the predicted slowdown of 80 milliseconds. The interaction may take the form of an augmentation of effect, in which case the slowdown for doubly drugged subjects would be greater than 80 milliseconds. Another possible form the interaction might take is that of a cancellation of effect, the drug user's fantasy that taking both drugs together will eliminate any deleterious effect of either.

The drug study is a 2×2 factorial design; a two-level first factor is paired with a two-level second factor. In all, there are four experimental conditions: control (neither alcohol nor marijuana), just marijuana, just alcohol, or both. Presenting the four conditions with these ordinary verbal labels hides the factorial nature of the design. However, we can see the structure by writing out the combinations, regarding each factor's levels as "administered" or "not administered." The four treatment combinations in the design could be represented as: 11 = first level of first factor, first level of second factor (i.e., both not administered); 21 = second level of first factor, first level of second factor (i.e., alcohol administered, marijuana not administered); 12 = first level of first factor, second level of second factor (i.e., alcohol not administered, marijuana administered); 22 = second level of first factor, second level of second factor (i.e., both administered). The dose for alcohol, when administered, is 50 ml, while the dose for marijuana, when administered, is 1 g. How do we assess the impact of the drugs?

The Factorial Plot

Interaction patterns can best be seen in graphs. A graph of cell means does not convey the significance information given by ANOVA, but it does show the direction and magnitude of the effects. This perspective is important not only for description, but also for inspiring theoretical explanation of the experimental results. One should form the habit of drawing such graphs, and it is worthwhile to practice describing various possible results. The plot is a fundamental part of the data analysis. After you have mastered the technique, you may avoid some of the labor by using the computer program FACTORIAL ANOVA.

The mean response goes on the vertical axis, along with an appropriate label. I sometimes omit the numbers on the axis, though it is usually a good idea to include them, and FACTORIAL ANOVA makes it easy to do so. Either factor may be placed on the horizontal axis, and the levels of the other factor distinguish the curves. The choice is not exactly arbitrary; sometimes it will be easier to understand the data from one perspective rather than the other. I'm reluctant to tell you to make two graphs, so I'll instead divulge my private, unofficial guideline. I usually prefer to put the factor with more levels on the horizontal, because that way fewer curves need to be drawn. The FACTORIAL ANOVA program includes an option to flip the roles of the factors back and forth, so it's easy to look at both orientations.

Each point, then, conveys the joint impact of one level of the first factor and one level of the second. It is helpful to use specific labels to denote the factor names and levels, rather than merely to refer to them with structural values (e.g., level 1, factor A).

Now we are ready to look at some hypothetical results for the drug experiment. Shown in each graph is a possible set of mean responses for each of the four treatment combinations.

The first figure on p. 33 shows the simplest pattern of results from an interpretive standpoint. Both treatment factors are effective (the RTs are higher when the drug is administered), and whether one drug has been administered or not does not affect the efficacy of the other drug. The latter phrase is synonymous with additivity, or lack of interaction, and its visual counterpart is parallel lines. The parallelism reflects the fact that the differences between the responses to the various levels of one factor are the same at all levels of the other factor.

FACTORIAL STRUCTURE   33

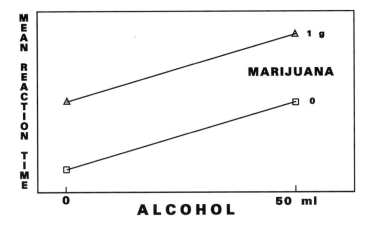

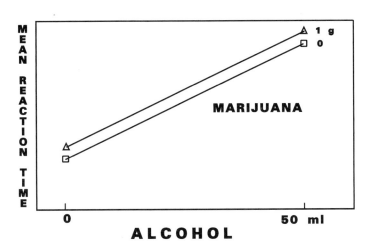

The lower figure also shows a simple pattern. Here too there is no interaction, as the lines appear parallel; but this time the presence or absence of marijuana seems to make little difference. In ANOVA language, we would say there is no MAIN EFFECT of the marijuana factor.

The other side of this coin is shown in the upper figure on p. 34. Again, there is no interaction; but this time it is the alcohol factor that is impotent, while the marijuana has what seems to be a sizable effect. Of course, the significance tests of ANOVA will be needed to furnish statistical evidence to buttress the visual impressions given in these graphs, because one cannot see the within-group (usually denoted as WITHIN-CELL in the factorial design, but it is the same term) variability.

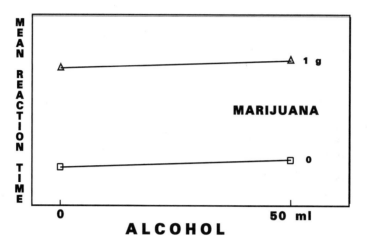

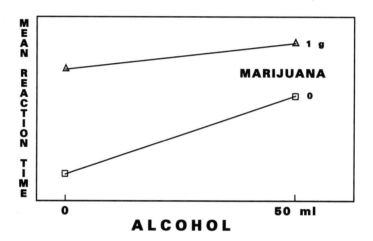

A decrease in impact is shown in the interaction depicted in the lower figure. The slowdown in reaction time produced by alcohol is reduced by the addition of marijuana; alternatively and equivalently, one could say that the slowdown associated with marijuana is reduced when alcohol has been administered as well.

The upper figure on p. 35 also shows interaction. Marijuana makes a greater difference when alcohol has been administered than when it has not. Of course, since interaction involves both factors equally, this figure may be summarized with the alternative statement that alcohol produces a greater differential effect when marijuana has been administered than when it hasn't. A creative researcher could base a theory of augmentation of drug effects on such results. Both drugs increase reaction time, and when combined their effects superadd.

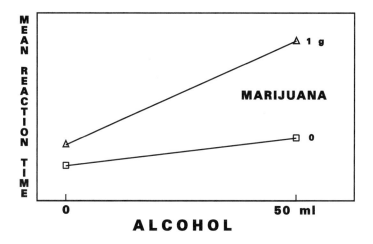

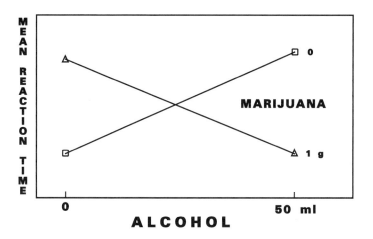

The pattern shown in the lower figure is called a crossover interaction for the obvious reason. The interpretation of this result would be that the deleterious effect of one drug is canceled by the addition of the second drug. Either drug by itself slows down the response, but both together restore its natural pace.

Range Limitation

As you scrutinize the plots, note that the values of the levels are given. This specification emphasizes that the observed pattern applies only to the levels used in the investigation. The pattern cannot be presumed to extend beyond the tested range. Although one may casually summarize results by asserting that factors interact, it should be kept in mind that this statement is a shortened version of the precise observation that interaction was found for the tested values. For quantitative variables, it may be reasonable to speculate about interpolated levels (values between those actually employed); drawing lines to join the points in effect does that. However, extrapolation to values outside the range of tested levels is definitely hazardous.

Explication and Explanation

When a research report is presented to a scientific audience, the investigator's responsibility (and pleasure) includes both explication and explanation of the results. Explication is the process by which the researcher guides the reader through the data, clarifying what was found in an objective manner. Explanation is an account of the reasons underlying the results; here is where creativity comes into play. Explication is localized in the results section of an APA-style paper, while explanation belongs in the discussion.

Algebraic Representation of Interaction

An algebraic counterpart of the two-way graph may comfort those who aren't visually oriented. Three statements, one for each factor and one for the interaction, furnish information equivalent to that in the graph. Each equation compares all of the cell means, the $\bar{x}_{ij}$, in a different way.

$$\text{first factor effect} = \bar{x}_{21} - \bar{x}_{11} + \bar{x}_{22} - \bar{x}_{12} \tag{4-1}$$

$$\text{second factor effect} = \bar{x}_{12} - \bar{x}_{11} + \bar{x}_{22} - \bar{x}_{21} \tag{4-2}$$

$$\text{interaction effect} = \bar{x}_{12} - \bar{x}_{11} - (\bar{x}_{22} - \bar{x}_{21})$$

$$= \bar{x}_{12} - \bar{x}_{11} - \bar{x}_{22} + \bar{x}_{21} \tag{4-3}$$

Assure yourself that these equations really do mirror what you can see in the graph. For each main effect, the terms that include the second level of that factor have plus signs attached to them, while the terms that include the first level of that factor have minus signs attached to them. Thus the comparison is between the sum of the means for each level of the factor, with the summation extending over both levels of the other factor. If the effect in question is nonexistent, then there should be no consistent difference between the means for its two levels, and thus the algebraic expression should yield a sum of zero. The interaction effect is an algebraic analogue of parallelism, as it examines two sets of differences. If these differences are of equal size, the subtraction will yield zero.

Factorial Computer Programs

While FACTORIAL ANOVA is quite clever in how it elicits data from the user, there is one element regarding your design of which the program is necessarily ignorant. The program user must specify the factorial structure: the number of factors and the number of levels per factor. If there are six scores, do you have a 3 × 2 design or a 2 × 3? The program doesn't care which factor is first and which second, but you might. The (arbitrary) rule used by these programs for data input is that the lowest-numbered factor cycles fastest. Therefore, it expects data from a 3 × 2 design in the order 11, 21, 31, 11, 22, 32. The program will label the cell indices as it prepares to accept the input values, so you don't have to memorize the pattern. But it is worth organizing the data layout prior to entry so that typing from your list of scores is maximally convenient. Determine how you want to align the data—going across the page in rows or down in columns, or any scheme you find comfortable—before you designate the factors. Choose as factor one the factor that cycles fastest in your preferred typing order. Select meaningful factor names (rather than accepting the program defaults of "A," "B," "C," and so on), so that you can remember what you did when you look at the output after some time as elapsed. Each factor's name must begin with a unique letter; this requirement is imposed to avoid confusion in reading the plots (and the tables that you'll learn about in the next chapter).

There is a trick you need to know. Like many programs designed to work with factorial designs, FACTORIAL ANOVA considers scores per cell as though it were a factor. Thus, there will be always be one factor more than you think there should be. In the factorial world, it is customary to refer to this additional factor as REPLICATES. If you have a 2 × 3 stimulus design with four scores per cell, the program must be told there are three factors. Because the program regards replicates as a factor, there must be an equal number of scores in each cell. (There is a theoretical reason for this limitation, discussed in the next chapter.) The replicates factor, with four levels, can be factor one, two, or three; the program doesn't care. For my style of presenting the hypothetical data comprising an exercise, with the scores in each cell bunched together, data entry will probably be easiest if you designate replicates as the first factor. The 2 × 3 stimulus design would be entered as a 4 × 2 × 3 design. When you enter the data, you will first put in the four responses to the first treatment combination. Next, move to the second level of factor two and enter the four responses to that combination. Having cycled through the second factor for the first time, go back to the first level of that second factor and move to the second level of the third factor; enter the four scores to that combination. FACTORIAL ANOVA will guide your entry by requesting the scores in the order consistent with the way you have designated the factors.

Using mnemonic labels for the factor names will help. You can also use mnemonic labels for the levels of each factor by checking the Use Labels for Levels option offered through the Preferences menu on the opening window. For the replicates factor, use cell indices rather than label names; for the other factors, meaningful labels will make it easier to keep track of what you are doing. After a little practice, you will become so proficient at data entry that you won't even need the cues presented by the program.

Here's the opening window for FACTORIAL ANOVA, set up for a two-factor design. Notice that I entered "3" as the number of factors, because of the replicates.

38 ANALYSIS OF VARIANCE AND FUNCTIONAL MEASUREMENT

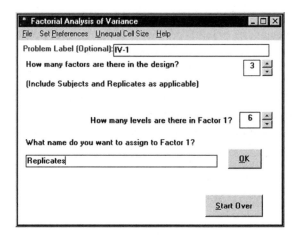

Here's the Data Entry window, as I am entering the data.

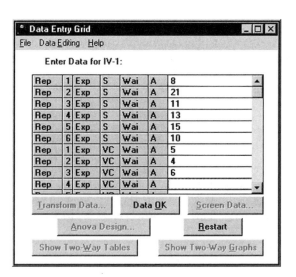

Clicking the Graph button produces a small window that asks which pair of factors should be displayed in the graph (or graphs). The Replicates factor is not of interest, so I chose to see the combination of the two substantive factors, Expectation and Waiting.

FACTORIAL STRUCTURE 39

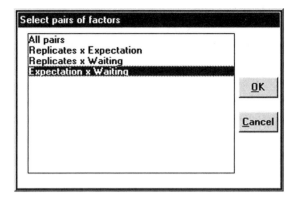

The graph can be edited in several ways; I have displayed the Y-Axis Options menu that allows possible modifications of the vertical axis, including changes of scale and positioning of the axis. Designating the minimum and maximum on the Y-axis is often advisable so that variation among the points may be seen clearly. If the default range chosen by the program is too large, all of the points are concentrated in a small area within the plot and so are difficult to distinguish. The Edit menu allows changes of the labels and fonts.

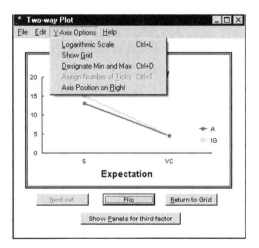

The Flip button interchanges the assignment of factors so that the factor currently occupying the horizontal axis becomes the factor that constitutes the various lines in the plot and vice versa. Either arrangement is acceptable; I usually put the factor with more levels on the horizontal axis because I think it looks better. The Show Panels button displays slices of the data at individual levels of a third factor. This option is useful only when there are more than two substantive factors.

The program produces plots in color, which is nice, but most scientific reports are restricted to black and white. I remove the color within Word by double-clicking on the pasted-in graph, then going through Format Picture to the Picture tab. Image Control needs to be set to black and white, with the brightness set at 38% and the contrast at 50%.

Exercises

Prepare factorial plots using cell means for the data presented below. It would be useful for you to do this both manually, at least for one or two problems, and with FACTORIAL ANOVA. Explore the program's graphing capabilities. Except for exercise 4-6, your graphs will look slightly different from those shown here because I used a specialized graphics program to polish most of the plots. Save the data from each problem in a mnemonically named file. You will be able to use that file to avoid some typing when you do the computational exercises for the next chapter.

Then, if you can find someone willing to listen, practice explicating the plots. You can only tell part of the story at this point, since the plot doesn't show variability. Significance questions are answered in the ANOVA table, which we'll explore in the next chapter.

4-1. A researcher measured effects of stress in terms of the amount of alcohol drunk by students who had free access to both alcohol and water. Volunteers were told that they would be in an experiment on learning in which wrong answers would be indicated either by painful electric shocks or by verbal correction. The students waited for the learning experiment either alone or in groups of four. While waiting, the subjects could drink as much alcohol or water as they wanted. The researcher's thought was that people anticipating shock would drink more alcohol, but those waiting alone would drink less than those waiting in a group. Describe the data. Notice that in this example, amount of alcohol drunk by an individual (measured in centiliters) is the dependent variable.

	Expectation	
Waiting	Shock	Verbal Correction
Alone	8, 21, 11, 13, 15, 10	5, 4, 6, 0, 2, 10
In group	20, 14, 22, 17, 3, 14	8, 0, 3, 5, 11, 0

4-2. A physiological psychologist studied the effects of nutrition and environment on the brain size of rats. The rats were fed a diet that either included vitamin

B_{13} or did not; the rats were allowed to play with either other rats, children, or graduate students for one hour per day. Postmortem analysis of the brain weights yielded the dependent variable, in grams. There were four rats in each experimental group. Describe the results.

	Environment		
Nutrition	Rats	Children	Grad Students
With B_{13}	21, 20, 21, 21	21, 22, 21, 21	24, 25, 25, 23
Without B_{13}	21, 23, 22, 21	22, 24, 25, 24	23, 22, 23, 25

4-3. To help me decide which course to write a textbook for, I decided to survey the reading behaviors of my students. I asked 5 randomly chosen A students and 5 randomly chosen C students in each of three classes to write the number of textbook pages read on the back of the anonymous evaluation forms they turn in at the end of the quarter. The primary question of interest is whether students in different classes read different amounts; I was also interested in the relation between student success and amount of reading.

	A	C
Introduction to Psychology	300, 247, 285, 300, 25	49, 58, 300, 75, 90
Intermediate Statistics	200, 235, 270, 210, 198	240, 250, 203, 190, 244
Advanced Statistics	310, 260, 310, 280, 170	50, 140, 190, 85, 310

4-4. A social psychologist conducted a study contrasting the sexual habits of high-anxiety and low-anxiety college students. The scores were the number of reported sexual experiences (sessions in which orgasm was achieved with a partner) during the previous month. There were five subjects in each group. Describe the data.

	High Anxiety	Low Anxiety
Males	4, 7, 3, 14, 10	12, 10, 15, 19, 11
Females	5, 5, 10, 6, 7	9, 5, 8, 10, 11

4-5. A therapist studied the effectiveness of three types of therapy for reducing smoking. He also thought that the length of the treatment might matter.

Twenty-four smokers were randomly assigned to six treatment groups. The scores are the number of cigarettes each participant reported having smoked during the week after the treatment had ended. Describe the impact of therapy type and duration.

	Duration	
Type of therapy	1 week	4 weeks
Willpower	140, 120, 98, 105	150, 132, 140, 160
Aversion	110, 92, 115, 126	0, 37, 15, 0
Drug	100, 95, 150, 135	91, 50, 56, 70

4-6. An educational psychologist examined the effects of busing and number of parents in household on the IQs of children. The children went to school either by bus or on foot, and they came from homes with either two parents, only a mother, or only a father. Five children in each classification were tested. Describe the effect of transportation and parenting in household on IQ scores.

	Transportation	
Parenting	Bus	Walking
Two parents	105, 97, 132, 110, 95	125, 108, 148, 89, 122
Mother only	88, 104, 101, 128, 112	117, 151, 125, 108, 106
Father only	104, 85, 96, 115, 130	132, 97, 110, 114, 103

4-7. Referring to the figures on pp. 33–35 displaying some possible results from a reaction-time study involving alcohol and marijuana, which outcome(s) would you root for (i.e., hope to see occur) if you were:

A whiskey distiller?
An advocate for the medical use of marijuana?
An officer in MADD (Mothers Against Drunk Driving)?
A hearty party hound?

Answers to Exercises

4-1.

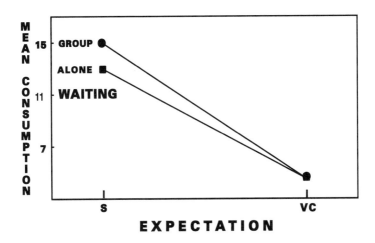

4-2.

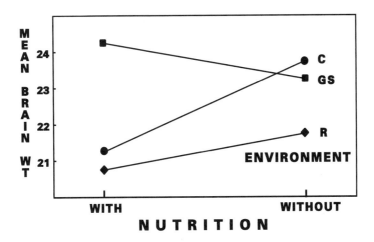

4-3.

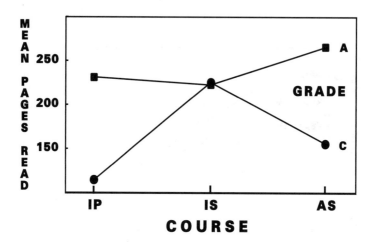

4-4.

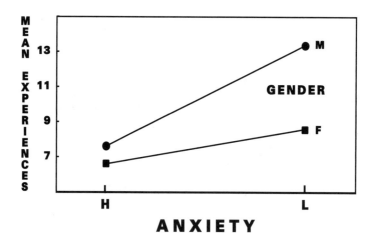

4-5.

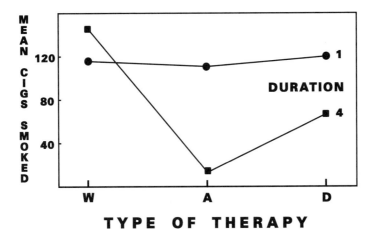

4-6.

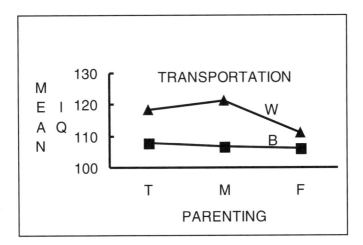

4-7. (1) The whisky distiller, hoping to show that alcohol consumption is harmless, would like the results to look like the upper figure on p. 34.

(2) The marijuana advocate would root for the lower figure on p. 33, wherein marijuana has little effect.

(3) The forces aligned against alcohol would like to see the figures on p. 33, where a dose of alcohol unequivocally slows down reaction time, or the lower

figure on p. 34, where alcohol slows the reaction, albeit less so when marijuana is also ingested.

(4) Those guys in the ΔΔΔ house who did any drug they could get their hands on would assure you that the lower figure on p. 35 will occur, since the two drugs taken together are harmless.

5

Two-Way ANOVA

A graph or a set of effect equations are not sufficient to answer the researcher's questions because of the variability issue. Statistical significance must be examined. In all of our factorial ANOVA problems, we shall have the same number of scores in each cell. Not only does this make for simplicity of computation and provide equal emphasis for all treatment combinations, but equal n also produces a desirable statistical property known as orthogonality. The import of this property is that the effects are estimated independently. In order to avoid the difficulties associated with a nonorthogonal analysis, most researchers routinely and resolutely aim for equal cell size. (Unequal n will be considered in chapter 12.)

To be concrete, let us consider an experiment on mathematics learning in children. The goal was to teach the children multiplication facts; two variables were of interest. First, what is the effect of the child being given a pocket calculator? Second, does the mode of instruction matter? Each child was assigned to either a group taught to use brute-force memorization, a group given insight into the nature of multiplication, or a group that learned a multiplying song. Training was carried out for a month, and then an exam was given. The dependent variable in the experiment was the child's exam score. A realistic attempt to answer these research questions would call for a sizable number of subjects, because children's scores are often highly variable. For illustrative purposes, six scores in each group will be given.

48 ANALYSIS OF VARIANCE AND FUNCTIONAL MEASUREMENT

Calcu-lator	Method		
	Memorize	Insight	Song
Yes	72, 83, 63, 48, 52, 90	27, 42, 64, 39, 72, 48	89, 94, 63, 99, 87, 92
No	92, 63, 45, 57, 59, 26	67, 55, 80, 22, 71, 66	91, 82, 78, 98, 100, 76

To get started, compute ΣX^2. As with one-way ANOVA, this number is simply the sum of the squares of all scores.

$$\Sigma X^2 = 183,436$$

Next, compute the total for each cell, and arrange the totals in an abbreviated table. All of the subsequent computations are based on the cell totals. Add horizontally and vertically to get the marginal totals. Although the illustration uses the abstract factor labels "A" and "B," when you work with data it is a good ideal to use specific labels such as "Calculator" and "Method." Similarly, in place of the numerical level labels "1," "2," and so on, it is advisable to substitute words denoting the treatments. The reason is simply that after a while, you will not remember what you meant when you inspect your computations. I'm sure you can tell how I know.

		B			
		B_1	B_2	B_3	
A	A_1	408	292	524	1224
	A_2	342	361	525	1228
		750	653	1049	

The sum of the cell totals is T, and from it you get T^2/N.

$$T^2/N = 2{,}452^2/36 = 167{,}008.44$$

Now we are ready to compute the sums of squares. Main effects are computed first. For each main effect, we need the marginal totals for each of its levels. These totals are squared, then summed; the sum is divided by the number of raw scores that went into each of the marginal totals. Afterward we subtract T^2/N.

$$SS_A = \frac{1}{18}(1{,}224^2 + 1{,}228^2) - T^2/N = 0.44$$

$$SS_B = \frac{1}{12}(750^2 + 653^2 + 1{,}049^2) - T^2/N = 7100.72$$

Notice that the divisor for SS_A is 18 because there were six raw scores in each of the three cell totals that comprise each of the two marginal totals. Similarly, the divisor for SS_B is 12, because six raw scores went into each of the two cell totals that went into each of the three marginal totals.

For the interaction, the cell totals are used. The cell totals are squared, summed, and then divided by the number of scores that comprise each total. Then the sums of squares for the main effects involved in the interaction are subtracted, as is T^2/N.

$$SS_{AB} = 1/6(408^2 + 292^2 + 524^2 + 342^2 + 361^2 + 525^2) \\ - T^2/N - SS_A - SS_B = 759.39$$

Here the divisor is six because there were six raw scores in each cell. Squaring totals and dividing by the number of scores that went into each total is a basic operation in ANOVA, one that appears again and again. A check that you have determined the divisor properly is available for all cases; if you multiply the divisor by the number of totals being squared within the adjacent parentheses, the product will be N, the number of scores in the experiment.

You may find it odd that no definitional formulas for the sums of squares are given for you to memorize. While it is certainly possible to write a set of expressions full of subscripts, my view is that it is easier to learn the computations using a template approach. The pattern is unvarying. For each source, I begin by writing

$$SS = \frac{1}{\ }(\ ^2 + \cdots + \ ^2) - T^2/N - \cdots$$

inserting as many plus signs and square symbols between the parentheses as there are relevant totals. Then I insert the subscript to the right of the SS and count the number of raw scores comprising each relevant total; that number goes under the division sign. After that, I subtract T^2/N. Finally, if the term being computed is an interaction, I also subtract the SSs associated with its components; in the example above, the components of the AB interaction are A and B.

Finally, the term for estimating the error variance is computed. In the factorial design, it is called the WITHIN-CELLS sum of squares. In fact, it is the same source referred to in the one-way case as the within-groups term.

$$SS_{WITHIN} = \Sigma X^2 - T^2/N - SS_A - SS_B - SS_{AB} = 8,567.00$$

This is a universal rule for computing the within cells sum of squares; subtract T^2/N and all substantive sources from ΣX^2. It is not difficult to verify the intuitively obvious idea that if there is only one score per cell, the within-cells sum of squares must be zero.

It is worthy of notice that the sum of the sums of squares for the three substantive sources, SS_A, SS_B, and SS_{AB}, is equal to what would have been computed as $\Sigma(t_j^2/n_j) - T^2/N$ had the six cells in the design simply been treated as though they were groups in a one-way ANOVA. The additional structure built into the experiment through use of a factorial design provides extra information by decomposing the groups. The factorial analysis, both for degrees of freedom and for sums of squares, is just a partitioning of the between groups sources.

All of the substantive sources are tested against the within cell term. Degrees of freedom are straightforward. For each factor, the degrees of freedom are one less than the number of levels for that factor ($df_A = 1$; $df_B = 2$). For an interaction, the degrees of freedom are obtained by multiplying the degrees of freedom for the factors involved ($df_{AB} = 2$). The within-cells df are computed as for a one-way ANOVA; subtract the number of experimental groupings (here $2 \cdot 3 = 6$) from the total number of scores ($df_{\text{WITHIN CELLS}} = 30$).

Construction of the ANOVA table completes the procedure:

ANOVA Table

Source	df	SS	MS	F
Calculator	1	0.44	$\dfrac{SS_A}{df_A} = 0.44$	<1
Methods	2	7100.72	$\dfrac{SS_B}{df_B} = 3550.36$	12.43*
Interaction	2	759.39	$\dfrac{SS_A}{df_A} = 379.69$	1.33
Within cells (error)	30	8567.00	$\dfrac{SS_{\text{WITHIN}}}{df_{\text{WITHIN}}} = 285.57$	

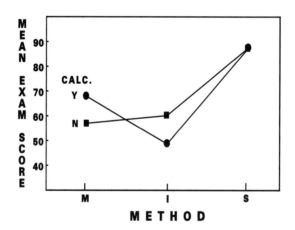

Interpreting the data entails considering the hypotheses built into the design. Examination of the cell means in the graph shown above confirms what the ANOVA says about main effects, namely that the teaching method had a dramatic effect, while the presence or absence of the calculator had none. As for the interaction, the nonparallelism suggested by the means was not sufficient to overcome the within-cell variability. The conclusion from these data is that teaching methods had differential effects on learning multiplication facts, but it did not matter whether or not the child had a pocket calculator available.

After you have done these calculations manually, use the FACTORIAL ANOVA program to reanalyze the data. Don't forget the trick about adding an "extra" Replicates factor.[1] The Design Specification screen lets you tell the program about the kind of analysis you want. For this analysis, the default options Fully Crossed and Independent Groups are appropriate; you need to click to tell the program which factor should be treated as the Replicates factor (even though it seems as though it ought to be obvious from the name I assigned—to the program, factor names are merely meaningless labels).

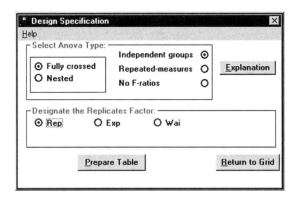

The Anatomy of Interaction

The sum of squares for the AB interaction depends only on the totals in the AB table. Let us now consider the question of how these totals for the education example would have to be different in order to produce a larger interaction. Recall the original table:

		B_1	B_2	B_3	
A	A_1	408	292	524	1224
	A_2	342	361	525	1228
		750	653	1049	

Suppose the data are modified by adding 20 to each of the six scores that went into the total for cell 1,3. Recall from chapter 2 that this will not change the error SS. Now the cell total will be increased by 120, so the AB table will be:

		B$_1$	B$_2$	B$_3$	
A	A$_1$	408	292	644	1344
	A$_2$	342	361	525	1228
		750	653	1169	

It should be intuitively clear that this modification of the scores in one cell will change not only the interaction SS, but will also affect SS_A, SS_B, and T^2/N. Only the within-cells term will be unchanged. Let us check our intuition by recomputing the ANOVA. The new data are identical to the old except that the six scores in the calculator + song condition are 109, 114, 83, 119, 107, and 112.

$$\Sigma X^2 = 206{,}796 \quad T^2/N = 183{,}755.11$$

$$SS_A = \frac{1}{18}(1{,}344^2 + 1{,}228^2) - T^2/N = 373.78$$

$$SS_B = \frac{1}{12}(750^2 + 653^2 + 1{,}169^2) - T^2/N = 12{,}534.06$$

$$SS_{AB} = \frac{1}{6}(408^2 + 292^2 + 644^2 + 342^2 + 361^2 + 525^2)$$
$$- SS_A - SS_B - T^2/N = 1{,}566.06$$

$$SS_{WITHIN} = \Sigma X^2 - SS_A - SS_B - SS_{AB} - T^2/N = 8{,}567.00$$

So SS_{AB} has indeed increased, but with the modification both main effects have increased as well. Is it possible to make a modification of the original data that will increase the interaction without at the same time changing the main effects?

The example we have just considered may suggest the answer to this puzzle. Obviously changing one cell will not work. Altering two cells will not work either (try it for yourself; you don't really have to do a complete ANOVA—work with the two-way table of totals). Two pairs of cell totals must be modified. For example, let's go back to the original scores and then add 20 to each score in cells 1,3 and 2,2. To maintain the same differences among the marginal means and thus keep the main effects constant, we must compensate by subtracting 20 from each score in cells 1,2 and 2,3. Let's look at the cell totals after this change:

	B			
	B_1	B_2	B_3	
A₁	408	172	644	1224
A₂	342	481	405	1228
	750	653	1049	

Since the marginal totals and T have the same values as in the original analysis, SS_A and SS_B will be what they were then. Although ΣX^2 is different with the new data ($\Sigma X^2 = 195{,}756$), the within-cells SS does not change. Only SS_{AB} is different in the ANOVA table; it is much larger ($SS_{AB} = 13{,}079.4$) because we built in an interaction.

Our anatomy lesson should clarify the nature of interaction. Interaction is concerned with differences between differences. That's why we had to change two pairs of cell totals to bring about a change whose sole locus was the interaction SS. Graphing the cell totals of the original and modified data should emphasize the idea that increased interaction is equivalent to increased nonparallelism.

It may be seen, then, that interaction is a characteristic of a pair of factors, not of a single factor. This structural property has an implication for the name of the source. The order of the contributing factors does not matter; thus, SS_{AB} and SS_{BA} are the same source. The freedom in assigning names also applies to the interactions of higher order (among more than two sources) we shall encounter in subsequent chapters.

Sums of Squares as Effects

There is a relation between the main effects as seen in factorial plots or the corresponding equations 4-1 and 4-2 and the sums of squares used in the ANOVA. This is as it should be, for obviously if an effect looks big to the eye it should impose itself in the formal statistical analysis. The relationship is easiest to demonstrate for a 2×2 design.

Consider the plot. What you see as an effect is a difference between cell means, visually averaged over all of the levels of the other factor. This difference may be translated into a quantity we'll call Δ, the difference between marginal means. There is a Δ for each factor. To make the idea concrete, let's consider a data matrix with two scores in each cell (note that it doesn't matter whether the horizontal factor is called A or B; the labels are arbitrary):

		A		Marginal	
		A_1	A_2	means	
B	B_1	5, 6	7, 9	6.75	$\Delta_B = 13.50 - 6.75 = 6.75$
	B_2	10, 14	12, 18	13.50	
		Marginal	means		
		8.75	11.50		
			$\Delta_A = 11.50 - 8.75 = 2.75$		

The quantity Δ is a direct measure of how potent is the effect; the larger it is, the more different is the effect of one level of the factor as opposed to the other. The sign of the difference is irrelevant in considering the magnitude of the effect. In terms of equations 4-1 and 4-2, each Δ is exactly twice the sum given by the corresponding equation.

Sums of squares and Δ values are related in a simple and dramatic way. The sum of squares for a factor is simply the square of the Δ for that factor, multiplied by the number of scores per cell. For the given data set,

$$SS_A = \frac{1}{4}(35^2 + 46^2) - 820.125 = 15.125$$
$$= 2(\Delta_A)^2 = 2 \cdot (2.75^2) = 15.125$$

$$SS_B = \frac{1}{4}(27^2 + 54^2) - 820.125 = 91.125$$
$$= 2(\Delta_B)^2 = 2 \cdot (6.75^2) = 91.125$$

Check this idea for yourself by trying an example with three scores in each cell.

The Δ derivation illustrates the fact that the sums of squares we use in ANOVA really do measure the magnitude of the effect. In extending the argument to larger designs, we must consider that there are progressively more Δ values as the number of levels of a factor increases. There are as many differences to consider as there are possible pairings of the levels; for three levels there are three pairings, for four levels there are six pairings. In general, for a factor with k levels, there are $_kC_2$ differences in marginal means to be squared and then summed. The notation $_kC_2$ is read as "the combination of k things taken 2 at a time";

$$_kC_2 = \frac{k!}{2!(k-2)!}$$

The notation k! is read not as "Kay!!!," but as "k factorial," which denotes the product of k and all of the integers below it. $k! = k(k-1)(k-2)\ldots 1$. For example,

$$_5C_2 = \frac{5 \cdot 4 \cdot 3 \cdot 2 \cdot 1}{(2 \cdot 1) \cdot (3 \cdot 2 \cdot 1)} = 10$$

For a two-factor design with r rows, c columns, and n scores in each cell, the row and column sums of squares may be expressed in terms of the effects as follows:

$$SS_{ROWS} = \frac{n \cdot c}{r} \Sigma(\Delta^2_{ROWS}) \qquad (5\text{-}1)$$

$$SS_{COLS} = \frac{n \cdot r}{c} \Sigma(\Delta^2_{COLS}) \qquad (5\text{-}2)$$

In these equations, $\Sigma(\Delta^2_{ROWS})$ is the sum of the squares of all of the differences between pairings of the row marginal means. $\Sigma(\Delta^2_{COLS})$ is a similar summation using the column marginal means.

Of course, equations 5-1 and 5-2 are not practical computational expressions for calculating sums of squares. The arithmetic using them is much more tedious than for the simpler expressions given previously. The point is to show that sums of squares can be given an intuitive interpretation in terms of differences between mean responses to the various levels of a factor. ANOVA looks at the square of what you see on the graph (plus variability).

Presenting the Results

ANOVA tables are but a subset of what is required for the analysis of a data set. For each significant difference, the analyst should also indicate which levels induce larger responses. Any important interaction should be described so that due caution in interpreting main effects may be exercised and to allow the construction of a theoretical explanation. The factorial plot is the usual way to convey this information. The significance level chosen by the researcher should be specified (and if it is a value other than .05, usually a justification for its use should be included). Inconsistencies between significance statements and apparent patterns among means should be discussed. To summarize, an analysis includes a table, a graph, and an explication.

For example, an explication of the results of the mathematics education study presented earlier in the chapter would guide the reader through the plot while incorporating significance information as follows: The apparent interaction between teaching method and the presence or absence of a calculator was not significant (at the .05 level). The method of instruction produced significant differences in exam scores, in that the song method generated results superior to those produced by memorization or insight. Whether a calculator was furnished, on the other hand, had no effect on performance.

A Caution

When interaction is present, interpretation of the main effects becomes difficult. The problem is that it is all too easy to forget about the interaction and carry away a conclusion about the main effect; yet that main effect, although significant, may attain its magnitude solely because of a particular, perhaps peculiar, combination of levels.

A conservative procedure for interpreting data is probably best. Always check for interaction first. If it proves nonsignificant, move on to the main effects. However, if the interaction proves significant, simplistic statements about the main effects are hazardous. A description of the main effects must incorporate the pattern of the interaction.

An Even More Worrisome Caution

Psychologists love to provide interesting explanations for interaction. Often overlooked is the possibility that an apparent interaction is merely an artifact, the result of a nonlinear response scale. In presenting the mathematics instruction study earlier, I glossed over potential difficulties with the response instrument. Suppose a true-false test had been employed to evaluate the impact of the instructional method. Even a student who learned nothing would be expected to score around 50%, while the most successful student could not top 100%. These limits might well constrict the range of scores so that true parallelism could not be seen. In providing a theoretical explanation for an interaction, the analyst should at least consider the possibility that it is spurious, especially if the points that appear out of alignment are at either extreme of the response range. Because the (dis)placement of even one mean can have dramatic effect on interaction, one must be confident that the response instrument is capable of providing measures linearly related to the true values before asserting the behavioral reality of an interaction. From whence cometh such confidence? See chapter 9 and especially chapter 14 for further discussion. Anderson (1961) and Busemeyer (1980) offer convincing illustrations of scale-dependent interactions.

The Shanteau Challenge

Suppose there were no FACTORIAL ANOVA program available. Could you do (most of) the calculations using the ONEWAY ANOVA program? Try it with the math education problem given above. The point of this exercise, suggested by James Shanteau, is to show that sums of squares can always be calculated using the one-way techniques. Of course, it's not very convenient to do factorial analyses this way. You'll have to make several passes through the data, and you won't be able to avoid using your calculator. But you will learn the important lesson that a program can be used for more than its primary purpose. Creatively deceiving a program is actually quite a valuable and legitimate thing to do. Sometimes there is no program designed to do exactly the analysis you want, but you can cope by understanding what the program you have is capable of doing.

Simple Effects

Because a factorial design may be viewed as a set of one-way designs, researchers sometimes elect to examine the one-way designs in isolation. The simple effects (sometimes called simple main effects) of a factor are the series of effects for a given factor at the individual levels of the other factor. Thus we may speak of the effect of the A variable at B_1, at B_2, and at B_3. In each case, the null hypothesis has the form "at B_i, the means for the various levels of A are equal." Similarly, we may look at the effect of the B variable at A_1 and at A_2. Each individual test provides the significance counterpart to what you see when you look at a single line in the factorial plot. From the perspective of either factor, if the simple effects are not the same at all levels of the other factor, then there is interaction. Thus, simple effects are a way to try to understand interaction. Keppel (1991) is a strong advocate of exploring simple effects after a significant interaction has been found.

Computation of the simple effects uses values from each individual row (or column) for the factor under examination. For the mathematics learning data given previously in this chapter, we have:

$$SS_{A \text{ at } B_1} = \frac{1}{6}(408^2 + 342^2) - \frac{750^2}{(6 \cdot 2)} = 363.00$$

It would be worthwhile to complete the table manually. Then use the Simple button made available by the FACTORIAL ANOVA program by returning to the Design Specification window after you have done an ordinary analysis.

Source	df	SS	MS	F
A at B_1	1	363.00	363.00	1.27
A at B_2	1	396.75	396.75	1.39
A at B_3	1	.08	.08	<1
B at A_1	2	4485.33	2242.67	7.85*
B at A_2	2	3374.78	1687.39	5.91*
Within cells	30	8567.00	285.57	

Although it is legitimate to calculate both sets of simple effects (as the program does), it is unlikely that a researcher would want to present both. The information presented is redundant. Notice that the degrees of freedom for the As and the Bs add up to more than the number of cells in the design; this is a sure sign that the sums of squares are not independent. In fact, the sum of the sums of squares for each source's simple effects add up to the sum of squares for that source plus the interaction. Thus, the sum of $363.00 + 396.75 + .08 = 759.83$, while (from the original ANOVA table) $0.44 + 759.39 = 759.83$. Similarly, $4,485.33 + 3,374.78 = 7,860.11$, while (again from the original ANOVA table) $7,100.72 + 759.39 = 7,860.11$.

Which set should be presented? In principle, the one that conveys the most information about the data, the one that has a natural interpretation. Generally, that will be the one with a larger effect. Here, that approach would favor examining B at the two levels of A. Another strategy researchers use is to report the simple effects for the variable with more levels; this has the consequence of looking at fewer effects. In the present case, one would again look at B at the two levels of A.

Finally, I might add another recommendation. It is not obligatory to examine or report simple effects. Do so only if you have a specific research hypothesis that implies that simple effects are important. One of the advantages of factorial

ANOVA is that a main effect provides generality precisely because it is averaged over all of the levels of the other factor.

More About Independence— The Single-Subject Design

The within-cells error term is appropriate for independent experimental groups. A primary assumption underlying the use of a common error term for testing all of the substantive sources is that the error variances in all of the cells of the design are comparable; accordingly, they may sensibly be pooled. For independent groups, this assumption is reasonable because differences between people, which comprise a major share of the variance, are likely to be spread equally across the groups.

Researchers also use the same analysis in a case in which the scores are distinctly nonindependent. In the SINGLE-SUBJECT design, one individual serves in all of the experimental conditions several times; that individual furnishes all of the data for the analysis. These repetitions (formally referred to as replicates, the same term we use in the independent groups context) for the one participant take the place of the several scores per cell that would come from different subjects in an independent groups design.

In a single-subject design, the scores cannot be independent because they come from the same person. This nonindependence need not bias the results if care is taken to avoid having the various responses interfere with one another (as would happen if, for instance, the observer remembered previous responses to similar stimulus combinations). The chief value of the design is reduction of variance because the personal contribution of the participant is a constant. Whereas in the ordinary within-groups design the personal contribution of the subject is confounded with the error component, in the single-subject design the personal contribution simply has no effect. In some other experimental settings, it is possible to estimate and extract the variance contributed by consistent interindividual differences. This issue will be explored when we consider repeated-measures designs in chapter 7.

The single-subject design is typically used when a good deal of training is required before stable responses are achieved. One such setting is psychophysics research, in which a person makes judgments about aspects of a set of perceptual stimuli. It is often difficult to learn to attend to the relevant stimulus dimensions, so the researcher wants to get a lot of data from each observer to justify the training expense. Usually a small number of observers is employed, each one generating several replications of the experimental design. The experimental hypotheses are tested on each subject's data separately. If similar results are obtained for most individuals, then the researcher can have some confidence in generalizing the conclusions to "people in general."

A recent application of the single-subject design is in the domain of expertise. Weiss and Shanteau (2003) have argued that two necessary characteristics of expert judgment are discrimination (responding differently to different stimuli) and consistency (responding similarly to similar stimuli). Expertise is assessed by asking the candidate to evaluate a set of stimuli presented one at a time. In order to get a measure of consistency, at least some stimuli must be presented more than once, and it is usual to employ a stimuli × repetitions factorial design. The authors combine the measures of discrimination and consistency in a ratio format called the CWS index. Although there are several possible ways to formulate the CWS index, in most cases it is computationally like an F ratio. The methodology has proven useful, especially in areas in which no obvious objective criterion of expertise is available. How can one determine whether a figure skating judge is evaluating the skaters correctly? Since there are no correct answers against which to compare the ratings, the CWS approach advocates examining judgments with regard to the fundamental properties of discrimination and consistency.

Exercises

Prepare ANOVA tables manually (calculatorially?) and with FACTORIAL ANOVA (if you didn't save the files from the previous set of exercises, regard this as an opportunity to practice your data entry skills). The problems are repeated for convenience; note the additional suggested exercises for problems 5-2 and 5-5. Integrate the tables with the plots to provide a full explication of the results; think of explication as an abbreviated version of the results section of a research paper.

5-1. A researcher measured effects of stress in terms of the amount of alcohol drunk by students who had free access to both alcohol and water. Volunteers were told that they would be in an experiment on learning in which wrong answers would be indicated either by painful electric shocks or by verbal correction. The students waited for the learning experiment either alone or in groups of four. While waiting, the subjects could drink as much alcohol or water as they wanted. The researcher's thought was that people anticipating shock would drink more alcohol, but those waiting alone would drink less than those waiting in a group. Evaluate the data. Notice that in this example, amount of alcohol drunk by an individual (measured in centiliters) is the dependent variable.

	Expectation	
Waiting	Shock	Verbal Correction
Alone	8, 21, 11, 13, 15, 10	5, 4, 6, 0, 2, 10
In group	20, 14, 22, 17, 3, 14	8, 0, 3, 5, 11, 0

5-2. A physiological psychologist studied the effects of nutrition and environment on the brain size of rats. The rats were fed a diet that either included vitamin B_{13} or did not; the rats were allowed to play with either other rats, children, or graduate students for one hour per day. Postmortem analysis of the brain weights yielded the dependent variable, in grams. There were four rats in each experimental group. Analyze the results. Then reanalyze the data after subtracting 20 from each score. When you carry out computations manually, the reanalysis should prove easier than the original analysis. Does this suggest a technique that might on occasion be useful? Making the numbers small is an old-timer's trick from precomputer days. When you repeat the process with FACTORIAL ANOVA, you can accomplish the subtraction instantly using the transformation option available in the Data Entry window.

	Environment		
Nutrition	Rats	Children	Grad Students
With B_{13}	21, 20, 21, 21	21, 22, 21, 21	24, 25, 25, 23
Without B_{13}	21, 23, 22, 21	22, 24, 25, 24	23, 22, 23, 25

5-3. To help me decide which course to write a textbook for, I decided to survey the reading behaviors of my students. I asked 5 randomly chosen A students and 5 randomly chosen C students in each of three classes to write the number of textbook pages read on the back of the anonymous evaluation forms they turn in at the end of the quarter. The primary question of interest is whether students in different classes read different amounts; I was also interested in the relation between student success and amount of reading.

	A	C
Introduction to Psychology	300, 247, 285, 300, 25	49, 58, 300, 75, 90
Intermediate Statistics	200, 235, 270, 210, 198	240, 250, 203, 190, 244
Advanced Statistics	310, 260, 310, 280, 170	50, 140, 190, 85, 310

5-4. A social psychologist conducted a study contrasting the sexual habits of high-anxiety and low-anxiety college students. The scores were the number of reported sexual experiences during the previous month. There were five subjects in each group. Analyze the data.

	High Anxiety	Low Anxiety
Males	4, 7, 3, 14, 10	12, 10, 15, 19, 11
Females	5, 5, 10, 6, 7	9, 5, 8, 10, 11

5-5. A therapist studied the effectiveness of three types of therapy for reducing smoking. He also thought that the length of the treatment might matter. Twenty-four smokers were randomly assigned to six treatment groups. The scores are the number of cigarettes each participant reported having smoked during the week after the treatment had ended. Analyze the impact of therapy type and duration. Examine the simple effects for therapy type.

	Duration	
Type of therapy	1 week	4 weeks
Willpower	140, 120, 98, 105	150, 132, 140, 160
Aversion	110, 92, 115, 126	0, 37, 15, 0
Drug	100, 95, 150, 135	91, 50, 56, 70

5-6. An educational psychologist examined the effects of busing and number of parents in household on the IQs of children. The children went to school either by bus or on foot, and they came from homes with either two parents, only a mother, or only a father. Five children in each classification were tested. Describe the effect of parenting and transportation on IQ scores.

	Transportation	
Parenting	Bus	Walking
Two parents	105, 97, 132, 110, 95	125, 108, 148, 89, 122
Mother only	88, 104, 101, 128, 112	117, 151, 125, 108, 106
Father only	104, 85, 96, 115, 130	132, 97, 110, 114, 103

Answers to Exercises

Although I won't include the factorial plots again (they are in the Answers to Exercises for chapter 4), incorporate what you see in them into your recapitulation of the results. I present one possible version of an explication for each problem. There are many acceptable variations in phrasing.

5-1.
Source	df	SS	MS	F
Expectation	1	541.50	541.50	22.24*
Waiting	1	6.00	6.00	<1
EW	1	6.00	6.00	<1
Within cells	20	487.00	24.35	

"People anticipating shock drink more than those anticipating verbal correction. Whether they wait alone or in a group does not matter."

5-2.	Source	df	SS	MS	F
	Environment	2	25.00	12.50	13.64*
	Nutrition	1	4.17	4.17	4.55*
	EN	2	12.33	6.17	6.73*
	Within cells	18	16.50	.92	

"Rats fed the vitamin develop smaller brains than those who don't get the vitamin, except if they play with graduate students. Rats who play with children develop larger brains than those who play with other rats, and those who play with graduate students develop even larger brains. However, among rats who don't get the vitamin, those who play with graduate students develop slightly smaller brains than those who play with children."

5-3.	Source	df	SS	MS	F
	Course	2	14024.07	7012.03	1.05
	Grade	1	42262.53	42262.53	6.32*
	CG	2	22782.07	11391.04	1.70
	Within cells	24	160496.80	6687.37	

"Students who received A grades reported reading more pages than those who received C grades, regardless of the course."

5-4.	Source	df	SS	MS	F
	Anxiety	1	76.05	76.05	7.04*
	Gender	1	42.05	42.05	3.89
	AG	1	18.05	18.05	1.67
	Within cells	16	172.80	10.80	

"High-anxiety students report fewer sexual experiences than low-anxiety students. Although males tend to report more sexual experiences than females, the difference is not significant."

Note that I added the sentence about gender to counteract the impression given by the graph.

5-5.	Source	df	SS	MS	F
	Therapy	2	18950.33	9475.17	27.75*
	Duration	1	9801.04	9801.04	28.71*
	TD	2	16750.33	8375.17	24.53*
	Within cells	18	6145.25	341.40	

Simple Effects	df	SS	MS	F
Therapy at 1 week	2	171.50	85.75	<1
Therapy at 4 weeks	2	35529.17	17764.58	52.03*
Within cells	18	6145.25	341.40	

"A four-week program is more effective in reducing smoking than a one-week program when aversion or drug therapy is used, but not when willpower is the therapy. For a one-week program, the type of therapy doesn't matter. However, for a four-week program, aversion therapy is the most effective, and drug therapy is intermediate in effectiveness, while willpower is the least effective."

The use of number of cigarettes smoked as a measure of treatment effectiveness is debatable. For discussion of considerations governing the choice of an appropriate index, see Weiss, Walker, and Hill (1988).

5-6.

Source	df	SS	MS	F
Transportation	1	780.30	780.30	2.72
Parenting	2	167.40	83.70	<1
TP	2	115.80	57.90	<1
Within cells	24	6893.20	287.22	

"Although children who walk to school tended to have higher IQ scores, neither transportation mode nor parenting in household significantly affected the IQ scores of the children."

As in exercise 5-5, I inserted the first clause of the summary statement to counteract the impression given by the graph. Omitting the clause would not be wrong, but including it is a good idea to help the audience.

Note

1. As you may have noticed, there's something slightly fishy here. Although we are treating replicates as a factor for the purposes of the program, it is an unusual factor. Because each subject is contributing only one score, the normal crossing relationship to the other factors exists only if we are willing to regard all of the first subjects in the various cells as though they were one entity, all of the second subjects in each cell as if they were a second entity, and so on. Technically, the structural relationship between replicates and the other factors is referred to as nesting; the replicates factor is nested under all of the others. The within-cells term we have been deriving manually by subtraction can be obtained more arduously using nesting computations. The complexities introduced by nesting are discussed fully in chapter 11, but we don't need to address them now. When you designate a factor as replicates in the FACTORIAL ANOVA program (the actual factor name doesn't matter—you designate by clicking), the proper algorithm is automatically invoked. Do *not* use the Nesting option offered by the program; choose the Fully Crossed option.

6

Multifactor Designs

One of the most attractive aspects of ANOVA is the fact that additional complexity does not necessarily mean new concepts must be learned. Three-way ANOVA simply calls for repeated applications of the computing rules used in two-way ANOVA. And four-way ANOVA is just repeated three-way ANOVA.

This simplicity is fortunate, for it is the analytical power of multifactor designs that makes ANOVA so valuable to the researcher. A careful investigator often wants to maintain experimental control over minor as well as major variables in the situation. For example, a complex study may call for several experimenters in order to get all of the subjects run quickly. These experimenters will inevitably be different in unknown, seemingly trivial ways no matter how well they are trained. By making experimenters a factor in the design, the researcher can pull out these differences in the form of a sum of squares for experimenters. Without such control, the interexperimenter differences would be lumped into the error sum of squares, thus weakening the power of the design to detect true differences brought about by the treatments.

Of course, the statistical analysis does not distinguish between substantively important variables and technically advisable control variables. In many experiments, all of the factors are of substantive interest. The limit on how many variables are of interest is the researcher's imagination. In planning the study, one prepares to give rational explanations of the expected main effects and interactions. As a practical consequence, the typical maximum on substantive variables is three, simply because it is difficult to conceptualize interactions of higher order. However, there have been many published reports of experiments with four

MULTIFACTOR DESIGNS

to six factors. To my knowledge, the record is 15 factors, held by Shanteau (1970). Prior to the era of the computer, carrying out an ANOVA of more than four factors was a major accomplishment, but nowadays it is a routine matter.

In order to execute the computations for a three-way ANOVA, we must reduce a three-dimensional data array to manageable terms. The first step is to present the scores in two-dimensional tables. Find the factor with the smallest number of levels, and let each level of that factor be one panel of the data display. For example, let us consider a $4 \times 2 \times 3$ design with two scores per cell. If we call the factors A, B, and C, respectively, the most convenient arrangement is to make two AC tables, one for each of the two levels of B.

As always, ΣX^2 and T^2/N are needed. For these data, $\Sigma X^2 = 4{,}377$ and $T^2/N = 3{,}870.02$.

B_1

	C_1	C_2	C_3
A_1	3,5	12,8	9,4
A_2	6,4	11,7	6,8
A_3	9,7	15,12	13,13
A_4	8,8	10,11	11,14

Raw scores

B_2

	C_1	C_2	C_3
A_1	5,10	3,6	12,9
A_2	7,8	4,7	9,7
A_3	10,9	9,7	16,12
A_4	6,11	10,14	15,11

To initiate the computations, we compress the raw scores into cell totals.

B_1

	C_1	C_2	C_3
A_1	8	20	13
A_2	10	18	14
A_3	16	27	26
A_4	16	21	25

Cell totals

B_2

	C_1	C_2	C_3
A_1	15	9	21
A_2	15	11	16
A_3	19	16	28
A_4	17	24	26

Next, we generate one of the two-factor tables, in which we add over the third factor, and then we make the other two two-factor tables. Order of construction is immaterial. Sums of squares are obtained from these two-way tables.

Arbitrarily we begin with the table of AC totals, and the sums of squares derived from it:

ANALYSIS OF VARIANCE AND FUNCTIONAL MEASUREMENT

	C_1	C_2	C_3	
A_1	23	29	34	86
A_2	25	29	30	84
A_3	35	43	54	132
A_4	33	45	51	129
	116	146	169	

$$SS_A = \frac{1}{12}(86^2 + 84^2 + 132^2 + 129^2) - T^2/N = 173.06$$

$$SS_C = \frac{1}{16}(116^2 + 146^2 + 169^2) - T^2/N = 88.29$$

$$SS_{AC} = \frac{1}{4}(23^2 + 25^2 + \cdots + 29^2 + 29^2 + \cdots + 54^2 + 51^2)$$
$$- T^2/N - SS_A - SS_C = 17.88$$

These computations are carried out just as in any two-way ANOVA; the divisors in each step are the number of raw scores that went into each of the totals in that computation.

Now we need to find the sums of squares that involve the other factor, B. These require both an AB table and a BC table. Arbitrarily we can construct the BC table first.

	C_1	C_2	C_3	
B_1	50	86	78	214
B_2	66	60	91	217
	116	146	169	

The elements in each cell are obtained by adding over the missing factor, that is, the total for $B_1C_1 = 3 + 5 + 6 + 4 + 9 + 7 + 8 + 8 = 50$.

$$SS_B = \frac{1}{24}(214^2 + 217^2) - T^2/N = 0.19$$

$$SS_{BC} = \frac{1}{8}(50^2 + 66^2 + 86^2 + 60^2 + 78^2 + 91^2) - T^2/N - SS_B - SS_C = 68.63$$

The last interaction, AB, requires the AB table.

	A_1	A_2	A_3	A_4	
B_1	41	42	69	62	214
B_2	45	42	63	67	217
	86	84	132	129	

$$SS_{AB} = \frac{1}{6}(41^2 + 45^2 + \cdots + 63^2 + 67^2) - T^2/N - SS_A - SS_B = 6.23$$

Finally, we come to the one new term in the three-way ANOVA, the three-way or triple interaction. SS_{ABC} is computed from the individual cell totals. In fact, a general rule in any multifactor analysis is that the highest-ordered interaction is computed from the individual cell totals. To get the sum of squares, all of the prior components are subtracted following what should by now be a familiar pattern.

$$SS_{ABC} = \frac{1}{2}(8^2 + 10^2 + \cdots + 28^2 + 26^2) - T^2/N - SS_A$$
$$- SS_B - SS_C - SS_{AB} - SS_{BC} - SS_{AC} = 39.23$$

Here the divisor is two, because there were two scores in each cell.

The error sum of squares is obtained by subtraction:

$$SS_{WITHIN} = \Sigma X^2 - T^2/N - SS_A - SS_B - SS_C$$
$$- SS_{AB} - SS_{BC} - SS_{AC} - SS_{ABC}$$
$$= 113.47$$

The rule given here for determining the error term, that is, subtract "everything" from ΣX^2, leads to somewhat redundant calculations, but it is certainly the easiest algorithm to remember. The within cells sum of squares for any multifactor design may be determined in this way.

With so many sums of squares to compute, it is easy to misplace one, so I recommend filling in the ANOVA table as you go along. Degrees of freedom for any interaction are computed by multiplying the degrees of freedom of the component elements.

Source	df	SS	MS	F
A	3	173.06	57.69	12.20*
B	1	0.19	0.19	<1
C	2	88.29	44.15	9.34*
AB	3	6.23	2.08	<1
BC	2	68.63	34.21	7.26*
AC	6	17.88	2.98	<1
ABC	6	39.23	6.54	1.38
Within cells	24	113.47	4.73	

Interpretation of Interactions

The display and interpretation of two-way interactions are straightforward, but the three-way interaction is another matter. A significant three-way interaction means that the two-factor interaction is different at the various levels of the third factor. There are three two-factor interactions in the three-way design, and the definitional statement about the triple interaction can be applied to all of them at once. That is, an ABC interaction means that the AB interaction is different at the different levels of C; or equivalently, the AC interaction is different at the different levels of B; or equivalently, the BC interaction is different at the different levels of A. One can regard the three-way interaction from any perspective; the same information about the data is conveyed.

Let us consider a situation in which we might expect a three-way interaction. Imagine students who have just completed a required introductory course in chemistry and who are now evaluating the quality of the instruction they have received. There were four different instructors. One gave amusing low-content lectures and easy exams; a second also gave amusing low-content lectures but had hard exams. A third instructor gave rich, fact-filled lectures and easy exams, while the fourth instructor gave rich, fact-filled lectures and hard exams. With this convenient factorial arrangement of faculty practices, we might (perhaps cynically) expect ratings of this sort:

	Lectures	
Exams	Low content	Fact filled
Hard	Low	High
Easy	High	Low

This interaction is what might be expected from typical students in an introductory course, but what about the chemistry majors and premeds? Forced to be more concerned with content, they might generate ratings like this:

	Lectures	
Exams	Low content	Fact filled
Hard	Low	High
Easy	Mediocre	High

These hypothetical results show a three-way interaction. The factors labeled "lectures" and "exams" interact differently depending on the level of the "student's major" factor. The triple interaction could equally well be expressed in a different way: the interaction in the ratings between student's major and lecture style depends on whether the exams are easy or hard.

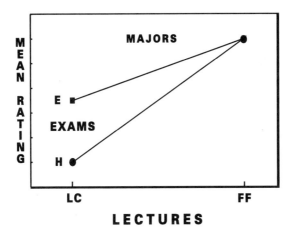

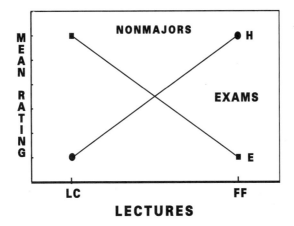

Unfortunately, the ease with which we can see two-factor interactions as nonparallelism on the factorial plot does not carry over to the case of three-factor interactions. The general statement is that a three-factor interaction implies the two-factor interactions look different at the various levels of the third factor. For example, if the upper figure were superimposed on the lower figure, the curves could not be aligned. The complexity of the three-way interaction is such that I do not deem it worthwhile to explore its anatomy, as was done in chapter 5 for the two-way interaction. There are too many forms the interaction may take.

I recommend proceeding mechanically when analyzing a three-way design. If the triple interaction should prove significant, then examine separate two-way panels as shown above in an attempt to describe and understand the complex pattern. Choose as the factor for each of whose levels you will construct a plot the one that will lead to the simplest descriptive statement. The Show Panels for third factor option in the FACTORIAL ANOVA program affords a convenient way to do this. It may also be worthwhile to look at standard two-way plots. This subdesign analysis entails extra effort but may be useful in diagnosing the locus of the three-way interaction.

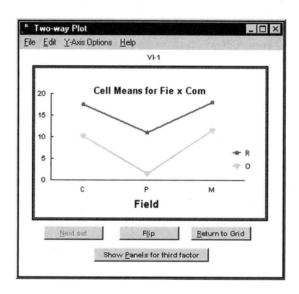

If the triple interaction is not significant, then move along to tests of each of the two-way interactions. The customary reservations about the acceptance of significant main effects when there are interactions involving these sources apply fully in this situation. The two-way interactions may complicate interpretations of the main effects; similarly, the three-way interaction may complicate interpretation of the two-way interactions. When a three-factor design is used, the researcher should be prepared to explain the complexity of the results that may emerge. In practice, researchers sometimes dismiss three-way interactions as Type I errors when no ready explanation is available, but such a strategy can hardly be recommended. Fixing the range so that the vertical axis for the various plots is always the same for a given data set, an option offered by the program, may facilitate comparisons across the plots.

High-way Designs

The computational formulas for designs with more than three factors follow naturally from what we have already seen. With a four-factor design, there are four three-way interactions to deal with, and each one is attacked separately in accord with what we have already seen. The four-factor interaction sum of squares is obtained by subtraction, using the individual cell totals as a base.

No examples of the four-factor design will be given, for two reasons. First, there are a lot of computational steps, and they are extremely repetitive; second, manually conducted four-factor ANOVAs have virtually gone the way of passenger pigeons. Designs with more than three factors are in my experience always analyzed on a computer.

Too Many Tests? Partial Analysis

The computer revolution has made it easy for researchers to incorporate many variables into their studies. Each additional factor not only brings increased experimental control and new knowledge, but also adds to the number of sources to be tested for significance. Designs with many factors have a great number of high-order interactions to be tested. For example, in a five-factor design, there are 16 three, four or five-way interactions. Anderson (1968) has suggested that for multifactor designs wherein some of the factors may be regarded as minor or technical, it may be advisable not to test the high-order interactions involving these less important factors. Such decisions should be made in advance of seeing the data; they should be based on knowledge of the experimental situation.

The point of Anderson's dramatic proposal (which should be read when you have had some experience with multifactor designs) is not to save labor. As we have seen, it can be difficult to describe high-order interactions cogently, let alone to explain them. Yet if a source unexpectedly proves significant, a conscientious researcher feels compelled to explain it. The argument is that it may be better not to test a source that would have little importance even if it were significant. The primary danger is that the more significance tests a researcher conducts, the more Type I errors are likely to occur. Carrying out a large number of significance tests on high-order interactions may well put a researcher in the awkward position of having to explain an interaction that is not real, even while she has no faith or interest in that interaction. It makes more sense to test only those sources about which one is eager to make a statement.

Evaluation of Partial Analysis

Partial analysis should not be engaged in lightly. It is all too easy to allow one's theoretical orientation to justify ignoring the data. On the other hand, routine significance testing in a multifactor experiment can mire the researcher in the quicksand of Type I errors. The more factors, the greater the likelihood of false positives. An alternative possible solution to the problem is to adjust the significance level for the various F tests. The Bonferroni procedure, discussed more fully in chapter 8, calls for dividing α by k, the number of intended tests (i.e., each test is carried out at the α/k level) in an attempt to control the experimentwise Type I error rate. The drawback in the case of factorial designs is that clearly some sources, such as main effects, are more important than others, and so one would generally wish to devote more statistical power to them. It might be possible to develop a scheme in which different significance levels are set in advance, on the basis of substantive theory, for the various sources in the ANOVA table. In effect, partial analysis is a qualitative approach to such a procedure.

Dependent F Tests

An associated difficulty with high-way designs is that one sampling error can lead to many wrong inferences. The problem is that the F tests conducted on the various sources in the design are not independent. Because the same denominator is used in all of the tests, if that denominator somehow emerges as too small, then there will be a host of Type I errors. In other words, if there is one spurious result, there are likely to be others. This awkward but unavoidable problem has been quantified by Hurlbut and Spiegel (1976).

Simple Effects in Multifactor Designs

Simple effects in higher-order designs work the same way that they do in two-factor designs, except that the arithmetic is a little harder if you do calculate manually. To find $SS_{A \text{ at } Bi}$, the respective squared sums must as usual be divided by the number of raw squares that comprise each sum, so the divisors must take into account the numbers of levels of the other factors C, D, and so on. Thus, for the example presented earlier in this chapter,

$$SS_{A \text{ at } B_1} = \frac{1}{2 \cdot 3}(41^2 + 42^2 + 69^2 + 62^2) - \frac{214^2}{4 \cdot 2 \cdot 3} = 100.16$$

Randomization Reemphasized

As we discussed in chapter 2, factors may be either experimental, that is, administered to subjects, or they may be classificatory, using characteristics the subjects have brought with them to the research. Classificatory variables, such as age, sex, or hair color, demand caution because they are the antithesis of random assignment. In a design employing both administered and classificatory factors, the researcher must make sure that subjects from each level of the classificatory factors are randomly assigned to the levels of the administered factors. Research employing only classificatory factors (e.g., the effects of age, sex, and race on IQ scores) runs the grave risk that unknown confounding variables bias the results. Indeed, such experiments, lacking the analytic power given by random assignment, can never yield unequivocal statements about causal relationships. The problem is the same one encountered with the use of the correlation coefficient as a research tool; correlation does not imply causation. From the technical perspective, ANOVA is insensitive to this distinction in the nature of the factors; all factors are analyzed in the same way. The researcher must keep this limitation in mind when planning and interpreting the study.

Exercises

6-1. As the students waited patiently in the bookstore's infinite checkout line on the first day of classes, the market researcher asked each a hypothetical question. "What is the top price you would pay for a (required, optional) (paperbound, hardbound) introductory text in (chemistry, psychology, music)?" Each student was asked one randomly determined combination of the three options and responded with an answer in dollars. The researcher collected answers from four students for each combination, interrogating a total of 48 respondents. Analyze these estimates.

	Paperbound				Hardbound		
	Field				Field		
	Chemistry	Psychology	Music		Chemistry	Psychology	Music
Required	11, 9, 10, 7	8, 6, 7, 8	12, 6, 10, 8	Required	25, 30, 23, 25	18, 15, 13, 13	30, 25, 25, 28
Optional	4, 5, 4, 6	0, 5, 2, 0	5, 7, 9, 9	Optional	20, 10, 18, 15	6, 0, 0, 0	15, 20, 12, 15

6-2. Ralph Raider researched the number of defects in new cars to see if there were patterns that could be used to guide buying recommendations. Cars were purchased from dealers either with or without the official dealer preparation. Six sedans and six hatchbacks were bought from each manufacturer. After delivery, Raider's mechanics counted the number of defects on each car. Analyze the data.

Farde sedans, with dealer preparation: 12, 14, 9
Farde sedans, without dealer preparation: 10, 15, 22
Farde hatchbacks, with dealer preparation: 15, 17, 19
Farde hatchbacks, without dealer preparation: 18, 22, 26
Dodger sedans, with dealer preparation: 16, 12, 18
Dodger sedans, without dealer preparation: 17, 22, 24
Dodger hatchbacks, with dealer preparation: 19, 24, 17
Dodger hatchbacks without dealer preparation: 23, 16, 14
Corsair sedans, with dealer preparation: 29, 33, 26
Corsair sedans, without dealer preparation: 34, 16, 15

Corsair hatchbacks, with dealer preparation: 37, 30, 25
Corsair hatchbacks, without dealer preparation: 40, 41, 36
Toytown sedans, with dealer preparation: 8, 3, 4
Toytown sedans, without dealer preparation: 3, 1, 2
Toytown hatchbacks, with dealer preparation: 0, 3, 2
Toytown hatchbacks, without dealer preparation: 2, 2, 3

6-3. In one of the most popular experiments I have conducted, male student volunteers (eager volunteers!) were given a free alcoholic drink and then timed as they solved a standard finger maze. Students were asked whether or not they were "regular" drinkers (usually had more than seven drinks per week). Twenty "regular" drinkers and twenty "nonregular" drinkers were signed on. Half of the participants were given a glass of milk to "coat the stomach" ten minutes before the alcoholic drink. The assigned drink was either a beer or a whisky highball, diluted so that both drinks contained the same amount of alcohol. Each volunteer was tested individually five minutes after finishing the drink. The dependent variable was the number of seconds it took to solve the maze. Five subjects were assigned to each cell of the design. Analyze the solution times.

	"Regular" Drinkers			"Nonregular" Drinkers	
	Beer	Highball		Beer	Highball
Milk	45, 28, 30, 64, 53	52, 39, 42, 77, 65	Milk	75, 93, 97, 83, 91	92, 105, 84, 78, 94
No milk	70, 42, 83, 98, 87	78, 67, 85, 73, 94	No milk	110, 86, 73, 59, 92	118, 133, 98, 76, 88

6-4. (This problem, with a large data set, is suitable only for the computer.) The data given below are from one of the subjects in a psychophysics experiment (Weiss, 1975). The subject's judgmental task was to choose the gray chip that appeared to be exactly intermediate in darkness between the two gray chips presented to him. This task is called bisection. The subject also engaged in trisection, in which the task was to find two chips that divided the interstimulus interval into thirds, and in quadrisection, whose definition you can infer. A response is the reflectance value (the percentage of light the chip reflects) of the gray chip chosen as the appropriate equisector. The higher the number, the lighter the chip. A value of zero would be perfectly black; a value of 100 would be perfectly white.

The stimulus pairs to be judged were assembled using a 3 × 3 factorial design, with all pairs of chips to be judged. Each pair was presented on a card. Thus, there were nine cards (three darkness levels of the left stimulus × three darkness levels of the right stimulus), each one of which looked something like this:

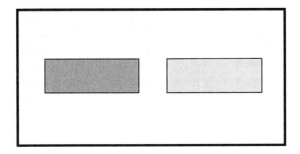

If we consider only the lightest responses from trisection and quadrisection, together with the bisection response, we have a 3 × 3 × 3 design (sectioning required × left stimulus × right stimulus).

The primary experimental hypothesis was that the two stimulus factors would be additive, that is, the interaction between them would be nonsignificant. It was assumed that all three main effects would be significant. It was of interest to determine whether the stimulus interaction, if it existed, would vary with section (so the three-way interaction was of interest). Evaluate these hypotheses for the one subject whose data are presented here.

Each subject went through four replications (in analyzing the data for this subject separately, we are employing a single-S design). The labels "L" and "R" refer to the location of the stimulus chips on the card containing a pair, and the levels "1," "2," and "3" refer to dark, medium, and light chips. The chips comprising the levels for the two factors were different.

		Bisection								
		Stimulus Combinations								
		L R	L R	L R	L R	L R	L R	L R	L R	L R
		1 1	2 1	3 1	1 2	2 2	3 2	1 3	2 3	3 3
Replication	1	4.61	10.43	19.77	12.00	24.58	43.06	15.57	39.55	68.40
	2	6.56	12.00	22.10	12.00	22.10	46.78	15.57	27.24	68.40
	3	4.61	12.00	17.60	13.71	24.58	43.06	15.57	27.24	59.10
	4	5.53	12.00	19.77	13.71	19.77	46.78	19.77	39.55	59.10

Trisection

		Stimulus Combinations								
		L R 1 1	L R 2 1	L R 3 1	L R 1 2	L R 2 2	L R 3 2	L R 1 3	L R 2 3	L R 3 3
Replication	1	3.82	12.00	19.77	7.72	24.58	36.20	10.43	33.05	63.64
	2	5.53	10.43	17.60	9.00	19.77	43.06	9.00	24.58	59.10
	3	5.53	12.00	17.60	10.43	24.58	43.06	10.43	36.20	63.64
	4	4.61	10.43	13.71	7.72	22.10	39.55	10.43	24.58	68.40

Quadrisection

		Stimulus Combinations								
		L R 1 1	L R 2 1	L R 3 1	L R 1 2	L R 2 2	L R 3 2	L R 1 3	L R 2 3	L R 3 3
Replication	1	3.82	10.43	15.57	6.56	24.58	39.55	10.43	27.24	59.10
	2	3.82	9.00	13.71	6.56	19.77	36.20	7.72	22.10	54.79
	3	4.61	10.43	12.00	7.72	24.58	39.55	7.72	27.24	59.10
	4	4.61	9.00	13.71	5.53	17.60	39.55	7.72	22.10	63.64

6-5. A breeder of laboratory rats conducted a study of some factors that might determine the size of a litter. Each female rat was assigned to a particular lab environment after she was weaned. The score was the number of pups in her first litter. The female rat lived in a cage with either one or two males. She was fed a diet of either rat chow, Wheaties, or mixed vegetables. The third independent variable in the research was the number of hours per day the lights were on. There were two female rats assigned to each treatment combination. Analyze the scores. Examine the simple effect of hours of light for each number of males.

	One Male					Two Males			
	Daily hours of light					Daily hours of light			
	4	10	14	17		4	10	14	17
Chow	7, 5	8, 4	4, 2	2, 3	Chow	2, 4	3, 2	1, 3	2, 2
Wheaties	10, 13	7, 9	6, 6	3, 5	Wheaties	1, 3	2, 2	3, 2	1, 2
Veggies	7, 6	5, 5	2, 3	4, 1	Veggies	2, 1	3, 2	1, 1	2, 3

Answers to Exercises

6-1.

Source	df	SS	MS	F
Field	2	688.79	344.40	51.61*
Compulsion	1	713.02	713.02	106.84*
Binding	1	1230.19	1230.19	184.34*
FC	2	17.79	8.89	1.33
FB	2	254.63	127.31	19.08*
CB	1	180.19	180.19	27.00*
FCB	2	10.13	5.06	<1
Within cells	36	240.25	6.67	

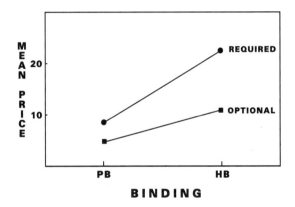

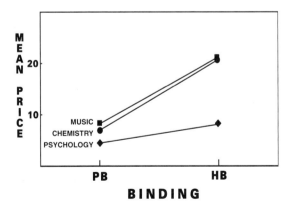

6-2.

Source	df	SS	MS	F
Prep	1	21.33	21.33	1.13
Model	1	154.08	154.08	8.19*
Brand	3	4546.17	1515.39	80.55*
PM	1	30.08	30.08	1.60
PB	3	52.17	17.39	<1
MB	3	217.42	72.47	3.85*
PMB	3	220.75	73.58	3.91*
Within cells	32	602.00	18.81	

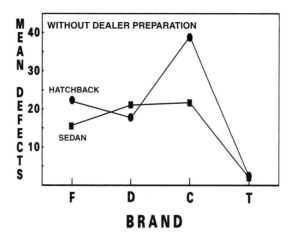

Because this problem has the most complex outcome in the set of exercises, I am including an explication. "Toytowns come with the fewest defects, while Corsairs have the most, and Fardes and Dodgers are in between. Hatchbacks have more defects than sedans, with the difference being largest for Corsairs. Dealer preparation doesn't matter, except that the difference between hatchbacks and sedans for Corsairs shows up only when there is no dealer preparation and vanishes when there is dealer preparation."

I selected the Show Panels for third factor option from the two-factor plot involving Model and Brand (not arbitrarily, because that two-factor interaction is significant, so I wanted to see it). Any of the two-factor plots will do for exploration of the three-way interaction. The view is different but the information is the same, namely that the two-factor plots look different at each level of the third factor. As Gertrude Stein might have said, "a three-way interaction is a three-way interaction is a three-way interaction."

6-3.

Source	df	SS	MS	F
Drink	1	801.02	801.02	2.99
Milk	1	2608.22	2608.22	9.75*
Regularity	1	7645.23	7645.23	28.57*
DM	1	42.02	42.02	<1
DR	1	30.63	30.63	<1
MR	1	1452.03	1452.03	5.43*
DMR	1	342.23	342.23	1.28
Within cells	32	8564.40	267.64	

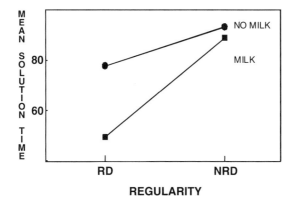

6-4.

Source	df	SS	MS	F
Section	2	434.23	217.12	26.87*
Left	2	18200.40	9100.22	1126.36*
Right	2	10032.30	5016.13	620.86*
SL	4	23.36	5.84	<1
SR	4	57.07	14.27	1.77
LR	4	4667.65	1166.91	144.43*
SLR	8	95.91	11.99	1.48
Within cells	81	654.42	8.08	

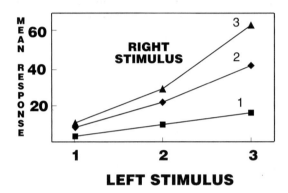

This experiment employed the single-subject design discussed in chapter 5.

6-5.

Source	df	SS	MS	F
Hours	3	54.06	18.02	12.18*
Diet	2	25.13	12.56	8.49*
Males	1	123.52	123.52	83.51*
HD	6	8.88	1.48	1.00
HM	3	41.56	13.85	9.37*
DM	2	28.29	14.15	9.56*
HDM	6	7.38	1.23	<1
Within cells	24	35.50	1.48	

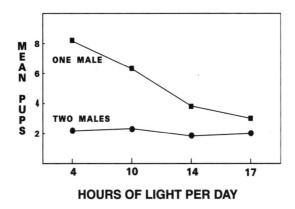

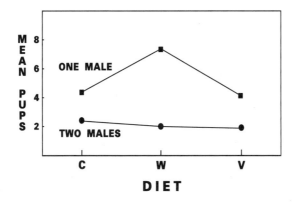

Simple Effects	df	SS	MS	F
Hours of light per day for 1 male	3	94.79	31.60	21.36*
Hours of light per day for 2 males	3	0.83	0.28	<1
Within cells	24	35.50	1.48	

7

Error-Purifying Designs

Repeated Measures

A repeated-measures design is one in which each of several subjects serves in more than one of the experimental conditions. In the simplest case, each subject gets all of the treatment combinations once. This new structure brings about two situations we have not encountered before. First, "subjects" comprises one of the factors in the design. Second, because there is only one score per cell, no within-cells term is available. (If each subject is in some but not all of the conditions, a repeated-measures nested design is being used—see chapter 11.)

The repeated-measures design has two important advantages for the researcher. One is that the design is economical. Once an experimental participant has been prepared or trained, it is obviously efficient to get more than one score from him. With human subjects, just getting them to the laboratory can be expensive. On the other hand, sometimes it is not feasible to get multiple observations from a subject because performance in one condition may affect performance in another condition. But when lack of independence between conditions is not a problem, the repeated-measures design can dramatically reduce the effort required of the experimenter.

The second advantage of the repeated-measures design is the reduction in error variance it affords compared to the independent-groups design. Because each participant contributes several scores, it is possible to isolate consistent differences among individuals. In the independent-groups design, idiosyncratic components are thrown into the error term along with other unsystematic variance. This mixing is inevitable because having only one score per volunteer does not allow the researcher to attribute anything particular to that individual. On the other hand, with the repeated-measures design, a sum of squares for subjects can be computed. This sum of squares is a main effect; it measures the differences among the totals for the various participants. The term is usually not of direct interest because few researchers are interested in demonstrating the truism that people (or rats) are different. Because the sum of squares has been computed, though, it can be thrown away to purify the estimate of error.

One should note, though, that the repeated-measures design is not simply an innocuous tool for reducing variability. It calls for a different experiment than one that explores the same variables but uses an independent-groups design. The subjects have a more complex set of tasks and may well respond differently. Grice (1966) has demonstrated that inferred behavioral regularities can depend on the researcher's choice of design even when identical values of the independent variable are employed.

In previous chapters there has been only one error term, a common denominator for all of the F ratios in an analysis. In a repeated-measures design, each substantive source has its own error term. The denominator of the F test for each source is the mean square for its interaction with subjects. Suppose there are two substantive, or treatment, factors, A and B. The experiment is established according to a three-factor design, $S \times A \times B$, where S indicates the subjects factor. Then the F ratio for testing the A main effect compares MS_A to MS_{SA}, the F ratio for testing B compares MS_B to MS_{SB}, while the test for the AB interaction compares MS_{AB} to MS_{SAB}. The actual computations proceed as for any three-factor design except that there is no within-cells sum of squares. The sense of these error terms is that when the mean square for a source is large relative to the mean square of that source's interaction with subjects (that is, relative to the differences between people with respect to that source), then the researcher can feel confident about generalizing to other subjects from the pool.

It is important to emphasize that no new formulas or computational schemes are called for in repeated-measures ANOVA. Subjects simply constitutes an additional factor. The arithmetic proceeds along familiar paths, with the number of scores per cell (n) fixed at 1. We have not previously encountered designs with one score per cell, but no changes in the steps leading to the sums of squares are needed. Only when F ratios are constructed is anything new called for.

To illustrate the repeated-measures analysis, let us consider a shot-putting experiment similar to the one described in chapter 1. The substantive variable in the research was monetary incentive; subjects were promised either $0, $1, or $10 per meter thrown. Each of six persons threw the shot under each of the incentive conditions. Here are the data:

	Incentive		
Subject	$0 per meter	$1 per meter	$10 per meter
A	20	23	30
B	50	50	60
C	37	48	59
D	15	17	21
E	54	56	62
F	8	11	15

The computations:

$$\Sigma X^2 = 28{,}844 \quad T^2/N = 636^2/18 = 22{,}472$$

$$SS_{\text{People}} = \frac{1}{3}(73^2 + 160^2 + \cdots + 34^2) - T^2/N = 5932.67$$

$$SS_{\text{Conditions}} = \frac{1}{6}(184^2 + 205^2 + 247^2) - T^2/N = 343$$

$$SS_{\text{PC}} = \frac{1}{1}(20^2 + 50^2 + 37^2 + \cdots + 62^2 + 15^2)$$
$$- T^2/N - SS_P - SS_C = 96.33$$

The ANOVA table is therefore:

Source	df	SS	MS	F
People	5	5932.67		
Conditions (Money)	2	343.00	171.50	17.80*
PC (error)	10	96.33	9.63	

Notice that only SS is given for people because the significance of that source would not be of interest here. The power of the repeated-measures analysis is shown by the significant F ratio for conditions. Had the same scores been collected from 18 separate subjects, the same conditions MS would have been swamped by the within-cells error MS (check this for yourself—the error MS would be 401.93).

The above analysis serves to illustrate the computations, but it is probably not adequate from an experimentalist's point of view. The difficulty is that each subject goes through the treatments in a specific order, and the ordering may have an effect on performance. A trial's residual impact may carry over to a succeeding trial. For example, fatigue may impede the last throw. Sometimes the problem may be alleviated by spacing trials over time, but that is not always practical. A careful researcher counterbalances order, to be sure. In this case, three balanced orderings were chosen, in the patterns 0, 1, 10; 1, 10, 0; and 10, 0, 1. Two subjects were randomly assigned to each. Therefore, order effects did not bias the estimate of the incentive variable. To get a balanced design, the number of subjects should be a multiple of the number of conditions. Counterbalancing should be routine practice when order effects are expected.

However, the effects of order are not eliminated by counterbalancing. Suppose, for example, that later trials produce larger responses. Those subjects who got a particular treatment later will score higher than they otherwise would have, and other subjects who were in that condition earlier in the experimental sequence will score lower. This will look like a treatment by subject interaction, and thus the order effect will find its way into the error term. Counterbalancing eliminates the possible bias in the estimation of the treatment effect, but it does not prevent inflation of the error. Because inflating the error term reduces the statistical power of the experiment, it is worthwhile to cleanse the error by pulling out the order effect. This entails expanding the experiment to make order a factor in the experimental design, and thus computing sums of squares for order and its interaction with incentive. Order effects may have empirical interest in their own right, of course—learning is the prime example[1]—but substantive concern is not the driving force underlying the extraction of these effects. To the extent that order really does play a role, incorporation of the additional factor into the analysis strengthens the conclusion about the primary incentive variable. However, if order actually has no effect, then using up degrees of freedom to analyze the source brings about a reduction, rather than an increase, in analytical power. That is why prior knowledge based on previous research or on pilot work is invaluable in planning statistical analysis even before data are collected. Order effects can either be averaged over or isolated for study, and in using a repeated-measures design, the researcher must make a choice. If it is already known that order doesn't matter much, then devoting experimental resources to account for its effects may be wasteful. In the absence of such knowledge, it may be worthwhile to examine order effects.

Additional Experimental Factors

Each substantive source requires its own error term, namely its interaction with the subjects factor. In a research report, the subjects source may not be labeled as such; the factor may have a name such as informants, or respondents, or patients, or rats, or sophomores, or any other term that identifies the group from whom the scores were collected. When the analysis is carried out using a program such as FACTORIAL ANOVA, the user must specify which term is the subjects source so that the program knows how to construct the appropriate denominators. Use of the repeated-measures design implies that the researcher is interested in the substantive sources and how they combine, not in individual differences among the subjects.

If there are two or more substantive sources, it may be of interest to dig into a pairwise interaction by looking at simple effects. The requirement that error terms be individualized for each source in a repeated-measures design applies to simple effects as well. The simple effect of A at B_i is calculated using the ordinary repeated-measures approach but with only the scores from the ith level of B. The heavy arithmetic load suggests using the program for these computations.

The Assumption of Additivity

The use of an interaction mean square as an estimate of error should strike you as questionable. In fact, this use is not particularly desirable, but it is a necessary compromise. The usual estimate of the random error component (σ_e^2), the within-cells mean square, is generally not available in a repeated-measures design because there is only one score per cell.

In a repeated-measures design, each of the substantive source mean squares contains three components: one is the effect of the source itself, the second is the random component, and the third is the interaction between the source and subjects. This interaction finds its way into the mean square, where it would seem not to belong, because of the peculiar nature of the subject factor. The subjects actually employed in a research project are viewed as a random sample from a larger population of potential subjects. If the source affects members of that population differentially (the definition of an interaction), then the apparent magnitude of that source will depend on which subjects have been chosen.[2]

As for the error terms, the mean square for the interaction of each substantive source with subjects includes two components—the random error component and the interaction itself. This means that the F ratio for a source compares {source plus random error plus subject-interaction} to {random error plus subject-interaction}. This is a sensible comparison, as the F ratio will be large if the source's contribution is large.

However, if the subject-interaction for a particular source is large, its presence in both numerator and denominator of the F ratio will dilute the comparison. That is why an important assumption of the repeated-measures design is that true subject-interactions are minimal. When this assumption is not satisfied, the statistical test has little power.

The additivity assumption can sometimes be tested; indeed, such testing can be an important facet of pilot work when a repeated-measures design is under consideration. If it is possible to administer each treatment combination two or more times to the subjects, then a within-cells mean square is available against which to test the source by subject interactions. The F ratios for these interactions should be nonsignificant if they are to be satisfactory error terms. Occasionally in such pilot work the F ratio for the source by subject interactions will be "too small," that is, close to zero; this is a warning that there is a hidden main effect inflating the within-cells term. This masked effect may be worthy of study in its own right.

Recognizing the Repeated-Measures Design

Correctly analyzing a repeated-measures design is not difficult, but the character of the design must be recognized. The key is seeing that each subject serves in more than one experimental condition. If each score comes from a separate subject, then we have an independent-groups design. The repeated-measures ANOVA is a generalization of the t test for correlated samples (sometimes called matched observations) and can be used whenever that t test would be used. The ANOVA is more general, since it can be used for any number of observations as well as for multifactor designs. Pretest versus posttest designs are common in some research domains and are often analyzed with t tests. But why should the researcher be limited to assessment at only two time points?

The repeated-measures analysis is effective whenever the scores from different conditions for the same subject are correlated, which is what the researcher usually expects. The assumption is that individual differences in underlying abilities will be maintained through the various treatment conditions; a "smart" subject will perform at high levels on the different cognitive tasks in one study, while a "strong" subject will perform at high levels on the different physical tasks in another. These consistencies make the analysis work in that consistent differences go into the sum of squares for subjects while inconsistencies increase the error sum of squares. When the sum of squares for subjects is large, the researcher should feel pleased that an independent-groups design was not used, because the differences between individuals would have made for a large within groups error term.

Blocking

Purifying the error term by extracting differences between subjects can also be accomplished when each subject serves in only one experimental condition. This is possible when a covariate, an additional score that measures some ability we think is related to performance in the experiment, is available for each subject. For example, suppose the research explores the number of words high school students recall three days after learning a list, with the variable of interest being the training method. Each student is randomly assigned to one of three training methods, and all are tested three days later. This is the sort of experiment in which one would expect considerable variability among people. With the analysis conducted according to an independent-groups design, the treatment effect might well be swamped by the within-cell variance.

One reason to expect individual differences is that a memory task is likely to reflect intellectual power; in general, high-IQ people remember better than low-IQ people. If IQ scores for the experimental subjects were available, it would be possible to isolate the effect of IQ by making it a factor in the design. This addition of a factor is the essence of blocking. The subjects are stratified, that is, artificially grouped by the researcher, so that differences in the covariate can be extracted. These differences will emerge as a main effect of blocks. The within groups sum of squares will be relatively small because the subjects within the various groups are homogeneous.

Blocking is a flexible and useful technique when a covariate is available.[3] Ideally, blocking of the subjects is carried out in advance; then within each block subjects are randomly assigned to conditions. This experimental design is known as randomized blocks. Computationally equivalent but somewhat less desirable is *post-hoc blocking*, in which stratification is performed after the data have been inspected (stratification on the basis of the covariate, not on the basis of the responses; the blocks are formed irrespective of condition membership). One cannot have as much confidence in a post-hoc analysis. A technical difficulty with post-hoc blocking is that it may not be possible to get enough subjects in the various blocks; with planned blocking and random assignment, achieving equal cell sizes is a routine matter.

While the computations for the blocked design proceed according to the ordinary ANOVA methods, creating the blocks allows scope for judgment. The first problem is to choose a sensible covariate for the particular experimental situation. When controlling error variability is the primary goal, the covariate will be effective if its correlation is in the range .2–.5. Correlations higher than .5 are infrequent in behavioral research, but it is easy to encounter logically connected variables that have correlations below .2. Fortunately, the blocked analysis is self-checking; if the correlation is too low the main effect of blocks is likely to prove nonsignificant. Researchers sometimes capitalize on this property of randomized block designs by using them in situations in which exploration of a person-related variable is the goal rather than increasing precision. For example, if an investigator hypothesizes that weight has an effect on emotional responses, she might randomly assign heavy and light subjects to the various anxiety-provoking situations that comprise the major treatment variable. Here blocks, or weight, would be a variable of substantive interest. Its correlation with the response would not be known in advance.

The second problem calling for judgment is the determination of the number and size of the blocks. Given the pool of available subjects, the goal is to select the number of blocking levels that yields maximum precision. The optimal number of levels depends on the correlation between the blocking variable (the covariate) and the response and on the number of treatment levels being administered. A table has been published (Feldt, 1958) giving the optimal number of blocks for various combinations of correlation, number of treatment levels, and size of subject pool. For example, if there are 30 subjects at hand, five treatments, and the correlation is thought to be .4, then the optimal number of blocks is 3. This means that the 30 subjects would be stratified into three groups of 10, and from each group two subjects would be randomly assigned to each of the five treatments.

The efficacy of blocking as a precision-increasing technique can be seen in the example below. Researchers in education compared two teaching methods.[4] Students from a Spanish I class were to learn a list of nonsense-word pairs after having been randomly assigned to one of the methods. The dependent variable was the number of trials required before the entire list was recited without error. There were 14 students assigned to each method. The ANOVA table:

Source	df	SS	MS	F
Methods	1	96.59	96.59	6.08*
Within-groups (error)	26	412.88	15.88	

This experiment actually was conducted in blocks because the researcher anticipated that the difference in methods would matter more with the academically poorer students than with the academically better ones. The cell means confirmed this idea:

		Method 1	Method 2
Grade in Spanish	A or B	8.86	6.86
	C or D	14.57	9.14

Fourteen A or B students and fourteen C or D students were used in the experiment. Since real-life classes rarely present such conveniently equal numbers, the researcher must randomly draw from the larger natural group (for example, there may be 18 A or B students) a number of subjects equal to the number in the smaller group. It is possible to conduct the blocked analysis with unequal group sizes, but it is more convenient to have equal n.

Compare the previous ANOVA with one in which the blocking factor is included:

Source	df	SS	MS	F
Methods	1	96.59	96.59	8.27*
Blocks (Grade)	1	112.01	112.01	9.59*
MB	1	20.56	20.56	1.76
Error	24	280.31	11.68	

Notice how the error term has been "purified" in this analysis. The error sum of squares in the previous analysis has been partitioned into three components (blocks, methods × blocks, and the reduced error). In this example, in which we have done both analyses, it is easy to see that the blocking has been effective; the F ratio for methods is higher. In everyday research life, only the blocked analysis would be performed. The effectiveness of the blocking is indicated by the significant F ratio for the blocking factor.

Considering the advantages blocking affords to the researcher, its cost is surprisingly low. There is effort involved in collecting the covariate scores and incorporating them into the design. From a statistical perspective, the cost is in degrees of freedom lost from the error term. When there are "enough" degrees of freedom (usually about 20 or more), this loss is no problem.

Exercises

7-1. Students were tested for ability to remember words under various drug conditions. Each of four students had to remember 10 words after drinking cola, acid, or water. Each student was administered a drug, tested, and then brought back the next day for a different drug. The drugs were administered to each student in a random order. Evaluate the drugs.

Student	Cola	Acid	Water
1	4	5	6
2	8	1	9
3	2	2	5
4	2	0	4

7-2. A graduate student recently completed a research project that he called "weight reduction clinic." "Not all of the patients lost weight," he reported, "but the program did cause significant weight change." Evaluate his claim.

Patient	Before	After
M	120 lbs.	115 lbs.
SF	195	187
HG	240	233
JL	180	182
DP	137	137
SR	192	177

7-3. In a study never published by industry sources, the time (in seconds) it took to drive a '55 Chevy one quarter of a mile was measured. Three different tire brands were used and four brands of gasoline. Three drivers were employed in the research; each drove the car under every combination of tire brand and gas. Analyze the data. Examine the simple effects of tire brand for each gasoline.

	Joe				Bob				Sam			
	Gas				Gas				Gas			
T	A	B	C	D	A	B	C	D	A	B	C	D
1	14	17	13	12	9	15	13	12	16	14	15	18
2	12	14	10	13	11	11	14	10	15	16	12	17
3	13	11	15	13	10	12	14	11	19	15	10	16

7-4. In a study with implications for driving performance, H. Moskowitz collected the times it took to respond to light flashes from 12 subjects. Each volunteer received four treatments in a programmed order. Treatment A was a placebo, treatment B was a dose of Benadryl, treatment C was a dose of alcohol, and treatment D was a combination of Benadryl and alcohol. Thus the treatments formed a 2×2 factorial design (each drug is either administered or is not). Order was not of experimental interest, but to control for its possible effects subjects 2, 6, and 9 got the order ABCD; subjects 7, 8, and 11 got the order BCDA; subjects 4, 5, and 12 got the order CDAB; subjects 1, 3, and 10 got the order DABC. Analyze the response times (in seconds); ignore possible order effects.

	Treatment			
Subject	A	B	C	D
1	1.476	1.737	1.863	4.381
2	2.503	2.698	3.891	3.460
3	2.869	4.482	4.768	4.875
4	2.786	2.830	2.862	2.607
5	2.501	3.036	2.534	4.414
6	4.585	8.014	3.676	5.533
7	4.594	6.317	5.732	5.525

| | Treatment | | | |
Subject	A	B	C	D
8	2.303	3.238	3.341	3.268
9	3.590	3.722	3.842	6.735
10	2.700	3.802	4.436	3.867
11	3.082	3.426	3.693	4.857
12	3.078	2.660	3.040	3.493

7-5. (a) Female subjects were told an experiment would involve paired-associate learning. One-third of the subjects were told that each correct answer would be worth $.05, one-third were told to "try as hard as you can," and the remaining third were told that each error would be punished with a painful electric shock. While waiting for the experiment to begin, subjects were invited to sit alone in the waiting room, in which there was a table with a bowl of crackers in it. Actually, the researcher was not concerned with learning, but with the effect of anxiety on eating, hypothesizing that subjects anticipating shock would eat more crackers. A colleague warned the researcher that this anxiety-eating relationship was likely to be much more pronounced in heavy subjects than in light subjects. Accordingly, when each subject arrived she filled out a "standard departmental questionnaire" that included, among other items, her weight. From this information, develop an assignment of subjects to the experimental conditions. Each subject's weight (in pounds) is given after her initials. GB (103), NW (128), EG (88), DC (133), RL (146), AE (102), SF (98), BK (127), LH (150), ZF (117), ER (108), SS (133), SR (120), AT (93), SC (96), EC (89), GG (129), KL (131), CW (110), DL (115), HH (148), NB (103), KR (105), JS (135), LN (132).

(b) Below is a portion of the data collected in the experiment described above. Analyze these data. What was the dependent variable?

| | Treatments | | |
Subjects	$.05	Try hard	Shock
Light	4, 2, 1	0, 3, 6	10, 8, 7
Heavy	5, 4, 9	8, 12, 5	20, 14, 22

7-6. Haralson and Richert (1979) explored aggressive behavior in fish as a function of feeding schedule. Each fish was tested at three drive levels (low, moderate, high), and the number of attacking responses was counted in each case. The researchers anticipated a connection between hunger and aggression. Because order effects might mask the relation, order was balanced over the six subjects. Analyze the data. What would be the next step?

ERROR-PURIFYING DESIGNS   93

Fish	Test order	Low	Moderate	High
1	MLH	60	81	37
2	HLM	81	61	94
3	MHL	5	9	6
4	LHM	15	2	7
5	HML	68	86	111
6	LMH	46	16	11

7-7. (This is another problem suitable only for a computer.) These data were collected by Weiss and Gardner (1979) while investigating an algebraic model of subjective hypotenuse estimation. Subjects saw a pair of lines of unequal length formed into a right angle. The task was to mentally construct a hypotenuse that would complete the triangle. Nine participants, all eighth-grade students who had not been trained in geometry, were run in the experiment; these data come from a practice condition. The two variables of interest were the orientation (the longer line was either horizontal or vertical) and separation (there were six possible differences in length between the two lines). Responses are the estimates of the length of an imaginary line connecting the two visible lines (in millimeters, produced by the subject's moving a sliding indicator). Analyze the data for the group as a whole.

					Separation				
		Orientation	1	2	3	4	5	6
Participant	1	H	235.8	361.0	323.2	293.2	362.0	334.0
		V	236.7	352.0	273.0	280.2	334.2	321.0
	2	H	198.0	350.5	318.3	300.5	361.0	352.2
		V	210.8	372.2	300.5	284.2	320.2	319.5
	3	H	138.5	378.0	273.2	238.7	342.5	314.8
		V	127.5	391.2	266.8	229.3	348.2	306.5
	4	H	156.0	399.3	316.3	303.3	400.3	390.3
		V	148.0	399.7	316.0	286.8	409.7	328.2
	5	H	185.5	449.3	345.3	310.7	413.2	388.7
		V	167.8	425.3	336.7	314.7	432.0	405.2

				Separation				
	Orientation	1	2	3	4	5	6
6	H	199.8	497.5	389.8	363.8	457.7	430.0
	V	199.2	499.2	427.0	336.0	479.2	425.2
7	H	145.0	388.0	280.0	261.7	373.0	359.8
	V	132.7	372.8	299.3	230.8	336.2	335.3
8	H	122.2	332.2	237.5	224.2	328.5	286.7
	V	107.0	312.3	218.0	201.7	295.5	258.0
9	H	107.3	309.3	259.0	254.0	323.5	322.0
	V	105.5	302.0	245.2	222.5	299.8	280.5

(Participant values: 6, 7, 8, 9)

Answers to Exercises

7-1.

Source	df	SS	MS	F
Students	3	30.00		
Drugs	2	32.00	16.00	4.36
SD (error)	6	22.00	3.67	

7-2.

Source	df	SS	MS	F
Patients	5	17827.42		
Treatment	1	90.75	90.75	4.89
PT (error)	5	92.75	18.55	

7-3.

Source	df	SS	MS	F
Drivers	2	71.72		
Tires	2	7.39	3.70	10.23*
DT (error)	4	1.44	.36	
Gas	3	5.00	1.67	<1
DG (error)	6	63.83	10.64	
TG	6	11.50	1.92	<1
DTG (error)	12	51.67	4.31	

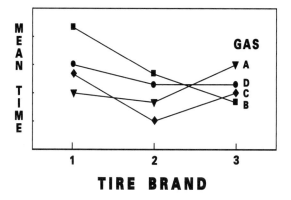

Simple Effects	df	SS	MS	F
Tires at Gas A	2	2.89	1.44	<1
DT at Gas A (error)	4	9.78	2.44	
Tires at Gas B	2	10.89	5.44	1.22
DT at Gas B (error)	4	17.78	4.44	
Tires at Gas C	2	4.22	2.11	<1
DT at Gas C (error)	4	21.78	5.44	
Tires at Gas D	2	0.89	0.44	<1
DT at Gas D (error)	4	3.78	.94	

7-4.

Source	df	SS	MS	F
Subjects	11	45.97		
Benadryl	1	7.71	7.71	11.90*
SB (error)	11	7.12	.65	
Alcohol	1	4.48	4.48	5.69*
SA (error)	11	8.65	.79	
BA	1	.01	.01	<1
SBA (error)	11	7.17	.65	

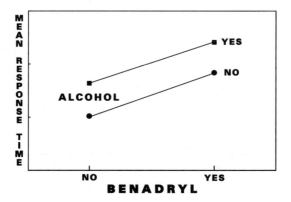

The factorial structure in this problem is hidden by the referral to treatments A, B, C, and D. Compare the design to that of the drug study used to illustrate factorial plotting in chapter IV. There is no source called order here, although with a more complex analysis (involving nesting, chapter 10) an effect might have been extracted. The researcher was not interested in the effect of order.

7-5. (a) With subjects blocked as suggested into two categories and three levels of the treatment factor, the number of subjects to be assigned should be a multiple of 6 (2·3). Since 25 subjects are available, it would be optimal to use 24 of them, with 4 subjects being assigned to each cell of the design. The simplest way to assign the subjects is to rank order their weights. Randomly assign the lightest twelve so that four go into each treatment, and do the same with the heaviest twelve. Subject ZF, whose weight is in the middle, will not be used.

(b) The dependent variable is the number of crackers eaten during the fifteen minute waiting period.

Source	df	SS	MS	F
Treatments	2	301.44	150.72	17.96*
Blocks	1	186.89	186.89	22.28*
TB	2	36.11	18.06	2.15
Within cells (error)	12	100.67	8.39	

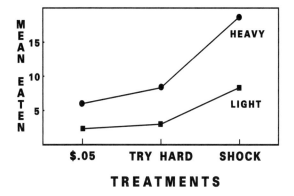

7-6.	Source	df	SS	MS	F
	Fish	5	19435.78		
	Drive	2	33.44	16.72	<1
	FD	10	3231.89	323.19	

Obviously the effect of drive is nonexistent here. Order effects might be worth systematic study; but perhaps the measure is too much under the control of unexplained, individualistic factors to reveal much.

I stopped indicating the error terms in the ANOVA table because by now you no longer need support. It is customary to indent them.

7-7.	Source	df	SS	MS	F
	Participants	8	204005.29		
	Orientation	1	3571.90	3571.90	13.31*
	PO	8	2147.06	268.38	
	Separation	5	587739.48	117547.90	117.59*
	PS	40	39985.53	999.64	
	OS	5	1193.32	238.66	1.41
	POS	40	6755.11	168.88	

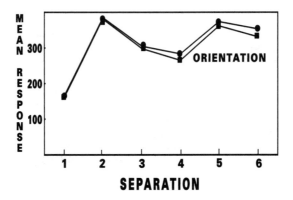

Notes

1. Poulton (1973) takes an extreme position on this issue, arguing that the repeated-measures design is empirically dangerous because subjects exposed to a series of stimuli are influenced by unwanted range effects. Responses are biased by the uncontrolled learning that takes place during the course of the experiment. In tasks with involving cognitive judgments, at least, this seems to me a concern of which the researcher should be aware. In my view, however, Poulton goes too far; it is possible to construct procedures that control for these biases. The simplest improvement is pretraining to expose subjects to the entire stimulus range prior to data collection.

2. The technical identification for an ordinary experimental variable is FIXED FACTOR (or FIXED EFFECT), while subjects is an example of a RANDOM FACTOR. The relevance of this distinction is the extent of generalization permitted. For a random factor, in which the levels have been selected randomly from the set of possible values, generalization to the entire set is allowable. Thus, it is customary to regard subjects as a random factor so that one may generalize beyond those chosen to those who might have been chosen as well. On the other hand, it is unlikely that the researcher might wish to choose stimulus levels randomly. Designated values constitute a fixed effect. The impact is that, for a fixed factor, generalization beyond the levels included in the design is not logically supportable.

3. Another technique that has been advocated when a correlated score is available is ANALYSIS OF COVARIANCE. Analysis of covariance has been out of favor among experimentalists; Keppel (1991, pp. 299–301) makes a strong case in favor of blocking. Maxwell, Delaney, and Dill (1984) attempt to revive ANCOVA. They suggest that it is more powerful and precise than blocking when covariate scores are available for all subjects prior to randomization and when the relationship between dependent variable and covariate is linear.

4. This example is from a research report by Pimsleur and Bonkowski (1961).

8

Specific Comparisons

Sometimes a researcher is interested in more specific information than whether all of the groups under study have the same mean. Of concern might be whether the mean of group 2 differs from the mean of group 5 or whether the average of groups 1 and 3 is equal to the average of groups 2 and 4. Specialized, single degree of freedom ANOVA tests known as specific comparisons are used to answer such questions.

Specific comparisons are planned by the researcher as either part or all of the data analysis. In general, planned comparisons should be viewed as an alternative to ordinary ANOVA rather than as a supplement. The choice of which comparisons to make must be made prior to inspection of the data because it is all too easy to convince oneself that the differences one sees are the differences one expected. The customary statistical inferences lose their validity when not based upon prior hypotheses. A post-hoc comparison option, discussed later in this chapter, is appropriate for confirming the significance of differences that came as a surprise.

When properly planned, specific comparisons can afford increased statistical power as compared to the overall ANOVA; the advantage is similar to that given by a magnifying glass in examining a small area. The specificity also makes a comparison more informative than an overall test, because it focuses on a particular question of central interest. On the other hand, if theoretical considerations or preliminary results have not suggested certain questions, then specific comparisons can be as useless as the same magnifying glass would be in examining a football field.

Carrying out a comparison simply involves constructing a sum of squares. The sum of squares is based on group totals and requires a set of cleverly chosen coefficients that are applied to those totals in a multiplicative way. These coefficients express the comparison. Deriving these coefficients is an easily learned art, perhaps best communicated by example.

Assume first that there are n scores per cell. This requirement of equal cell size is not absolute, and the computational algorithm can be generalized, but in research practice, equality is the rule. Let t_j be the total for the jth cell, and let C_j be the coefficient for that jth cell. Then if the algebraic sum of all of the C_js is zero, the sum of squares for the comparison is given by:

$$SS_{Comp} = \frac{(\Sigma C_j t_j)^2}{n \cdot \Sigma C_j^2} \tag{8-1}$$

This sum of squares is also a mean square, since each comparison is defined on one degree of freedom.

For an example of how the coefficients are determined, let's suppose that there are four groups in the experiment, and the comparison of major interest is between groups 1 and 2. The coefficients should be set up to contrast groups 1 and 2 and to ignore the others; thus an appropriate set could be $C_1 = 1$; $C_2 = -1$; $C_3 = 0$; $C_4 = 0$. As required, the sum of the C_js is zero. Equally acceptable would be the set $C_1 = -1$; $C_2 = 1$; $C_3 = 0$; $C_4 = 0$. The nonzero coefficients are chosen so that if the null hypothesis (the groups of interest have the same mean) is true, then the numerator of the sum of squares for the comparison will be zero. One could as well use the coefficient set $C_1 = 2$; $C_2 = -2$; $C_3 = 0$; $C_4 = 0$. Using twos instead of ones simply makes the arithmetic slightly harder, so it is customary to use the smallest integers possible. The sum of squares can be seen to express the contrast:

$$SS_{Comp} = \frac{[1 \cdot t_1 + (-1) \cdot t_2]^2}{n \cdot [(1)^2 + (-1)^2]} = \frac{1 \cdot (t_1 - t_2)^2}{2n}$$

Next, we consider an example in which two distinct comparisons are made. There are three groups: one experimental group, one normal control group, and one added control group that is expected to be equivalent to the normal control group. There are two obvious hypotheses, and these correspond to the two comparisons. The first comparison contrasts the two control groups, and the second compares the experimental group to the average of the control groups. It is convenient to list the coefficients in a table whenever there is more than one comparison:

	Experimental	Normal control	Added control
Comparison 1	0	1	−1
Comparison 2	2	−1	−1

Since the sum of squares for each comparison has been computed, significance tests can be carried out using the ordinary F table. The customary error term for all of the comparisons is the within-groups term. On rare occasions when the variances of the different cells of the design are extremely heterogeneous, researchers have been known to construct within-cell terms based only on the cells involved (that is, those having nonzero coefficients) in each particular comparison.

The answers to specific questions are often of more interest than whether treatments are differentially effective. For example, consider a study of pain-reliever effectiveness in which people were given a coded capsule to be taken when the next headache occurred. The volunteers were to record the number of minutes between taking the pill and subsequent absence of pain. Here are the results reported by the participants, who did not know which drug they had received:

Aspirin	Blufferin	Ansin	Tyrolin	Codeine
31	37	27	53	10
23	28	39	28	15
42	27	26	17	13
39	53	29	25	22
$t_1 = 135$	$t_2 = 145$	$t_3 = 121$	$t_4 = 123$	$t_5 = 60$

The first question of interest is whether the codeine is more effective than the less potent drugs. The respective coefficients for this comparison are 1, 1, 1, 1, and −4. The sum of squares for this first comparison is therefore:

$$SS_{\text{Comp1}} = \frac{(1 \cdot 135 + 1 \cdot 145 + 1 \cdot 121 + 1 \cdot 123 + (-4) \cdot 60)^2}{4 \cdot [1^2 + 1^2 + 1^2 + 1^2 + (-4)^2]}$$

$$= 80{,}656/80 = 1{,}008.2$$

The sum of squares for error is computed in the customary way for an independent-groups design. The error term for this comparison will be used for the others as well. It has 15 df.

$$SS_{\text{Error}} = \Sigma X^2 - \Sigma(t_j^2/n_j)$$
$$= 19{,}718 - 18{,}155 = 1{,}563$$

The F ratio for this first comparison is thus:

$$\frac{MS_{\text{Comp1}}}{MS_{\text{Error}}} = \frac{1{,}008.2}{104.2} = 9.68 *$$

confirming the researcher's suspicion that the more powerful drug was indeed more effective than the others.

102   ANALYSIS OF VARIANCE AND FUNCTIONAL MEASUREMENT

Also of interest is whether aspirin proved more effective than its derivatives, blufferin and ansin. The coefficients for this second comparison are −2, 1, 1, 0, and 0. The sum of squares is:

$$SS_{Comp2} = \frac{(-2 \cdot 135 + 1 \cdot 145 + 1 \cdot 121 + 0 + 0)^2}{4 \cdot [-2^2 + 1^2 + 1^2 + 0^2 + 0^2]} = 0.667$$

The F ratio for the second comparison is:

$$\frac{MS_{Comp2}}{MS_{Error}} = \frac{0.67}{104.2} \quad \text{which is} <1.$$

The third planned comparison is between the two aspirin modifications, blufferin and ansin. The coefficients for this comparison are 0, 1, −1, 0, 0. The sum of squares is:

$$SS_{Comp3} = \frac{(0 + 1 \cdot 145 - 1 \cdot 121 + 0 + 0)^2}{4 \cdot [0^2 + 1^2 + (-1)^2 + 0^2 + 0^2]} = 72$$

The F ratio for the third comparison is also <1:

$$\frac{MS_{Comp3}}{MS_{Error}} = \frac{72}{104.2}$$

Specific comparisons can be carried out with the computer program COMPARISONS. The input framework will be familiar; it is similar to that of ONEWAY. The comparisons you wish to do are conveyed in the Set up Comparisons window. The program user must determine the coefficients corresponding to each comparison. Here, I've just entered the coefficient assigned to the codeine group for the first comparison.

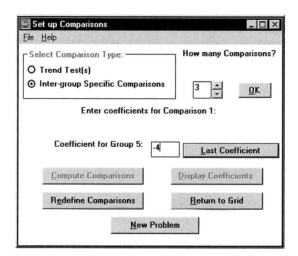

An imaginative researcher can find many comparisons that seem interesting. A reasonable limitation on the number of permitted comparisons is the number of degrees of freedom for substantive sources (that is, one less than the number of groups). The reason for this limit is that it is desirable to keep the questions asked of the data independent of one another.[1] Each comparison should produce new information. Independent questions are characterized as orthogonal. The maximum possible number of orthogonal comparisons is equal to the number of degrees of freedom.

It is not always easy to determine on an intuitive basis whether a set of comparisons is orthogonal. The key idea is whether the answer to one question determines to some extent the answer to another. This can be a tricky issue. For instance, consider an experiment with four groups. Here are four possible comparisons phrased as questions. We know that at most three can be orthogonal. Test your intuition on various combinations.

(1) Does group 1 differ from the average of groups 2, 3, and 4?
(2) Does group 2 differ from the average of groups 3 and 4?
(3) Does group 3 differ from group 4?
(4) Does the average of groups 1 and 2 differ from the average of groups 3 and 4?

My intuition in this context is unreliable, but fortunately there is a simple algorithm for determining orthogonality. Two comparisons are orthogonal if the sum of the products of their corresponding coefficients is zero. To apply this rule, list the coefficients for the comparisons in a table:

Comparison	Groups			
	1	2	3	4
(1)	−3	1	1	1
(2)	0	−2	1	1
(3)	0	0	1	−1
(4)	1	1	−1	−1

First we shall check on whether comparisons (1) and (2) are orthogonal. Expressed algebraically, the rule is:

$$\sum_j C_{\text{Comp1}j} \cdot C_{\text{Comp2}j} = (-3) \cdot 0 + 1 \cdot (-2) + 1 \cdot 1 + 1 \cdot 1 = 0$$

The coefficients are taken from the table and multiplied, and then the products are summed. The resulting zero means that comparison (1) and (2) are orthogonal.

What about comparisons (2) and (3)?

$$\sum_j C_{\text{Comp}2j} \cdot C_{\text{Comp}3j} = 0 \cdot 0 + (-2) \cdot 0 + 1 \cdot 1 + 1 \cdot (-1) = 0$$

So comparisons (2) and (3) are orthogonal. Comparisons (1) and (3) are as well. Comparisons (1) and (4) are not:

$$\sum_j C_{\text{Comp}1j} \cdot C_{\text{Comp}4j} = (-3) \cdot 1 + 1 \cdot 1 + 1 \cdot (-1) + 1 \cdot (-1) = -4$$

It should not be surprising that all four comparisons cannot be orthogonal. Each comparison uses up one *df*, and only three *df* are available with four groups. Many sets of orthogonal comparisons can be defined on four groups, but the researcher need not use up all three *df*. Comparisons should be planned only when the questions are interesting from a research perspective. An experiment with four groups could call for one, two, or three comparisons to be made. If two or three comparisons are planned, normally they should be set up so that all pairs of comparisons are orthogonal. When all of the degrees of freedom are consumed in orthogonal comparisons, the sums of squares for all of the comparisons will be equal to the conditions sum of squares. COMPARISONS checks the validity of each entered set of coefficients, and also checks that the comparisons are orthogonal.

Multiple Comparison Issues

When a researcher decides to perform a large number of F tests, an obvious concern is that spuriously significant F ratios will be observed just because Type I errors are expected to occur $\alpha \times 100\%$ of the time. Indeed, if k tests are carried out, the probability of one or more differences emerging as significant even though the null hypothesis is true in every case is $1 - (1 - \alpha)^k$. This quantity is roughly equal to αk. The exact probability can be determined using the program ALPHAK, from which output using the default settings is shown below.

Probabilities — Probability of at least one Type I Error		
Number of tests / Sig. Level	.05	.01
2	.098	.020
3	.143	.030
4	.185	.039
5	.226	.049
6	.265	.059
7	.302	.068
8	.337	.077
9	.370	.086
10	.401	.096
11	.431	.105
12	.460	.114
13	.487	.122

A simple way to cope with inflation in the error rate is the Bonferroni procedure, in which the significance level used for each test is α/k. The employment of this procedure limits the experimentwise probability of a Type I error to α. The down side of this decision is that the power for each individual comparison is consequently reduced. Thus, the choice of whether or not routinely to employ the Bonferroni procedure depends on one's philosophy regarding the trade-off between Type I and Type II errors. The traditional view, espoused eloquently by Petrinovich and Hardyck (1969), is that it is "better to punish truth than to let falsehood gain respectability"; because Type I errors become public, they are the chief worry. On the other hand, Davis and Gaito (1984) have suggested that the cumulative nature of research provides sufficient protection against the proliferation of Type I errors. They conclude that Type II errors, in which possibly promising ideas get buried by overly strict statistical procedures, are the greater danger in science. The appeal of this alternative view is that a researcher who adds additional conditions to a study ought to be rewarded, rather than punished, for diligence. The experimentwise approach makes it less likely than a given outcome will yield significance for the researcher who has gone the extra mile.

Trend Tests

A subset of specific comparisons is of special interest. This subset is concerned not with comparing groups or combinations of groups but rather with looking for specific trends in the group means. Such trends can be sought only when the independent variable is quantitative, that is, numerical. The researcher may predict that the means can be described by a straight line; an alternative phrasing is that the data follow a linear trend. This prediction of linearity is a fairly common one. Less frequently, researchers test for a predicted quadratic trend, or even for a cubic trend. The tests are carried out in the same way as are the other planned comparisons, except that the coefficients have traditionally been obtained from a table. The tabled values are known as orthogonal polynomial coefficients, a label that is accurate but not particularly precise, as other sets of coefficients not concerned with trends may also be orthogonal. As discussed below, I prefer to get coefficients from the computer program COMPARISONS, although for linear comparisons the coefficients are easy enough to construct in one's head.

Tabled coefficients may be used when the values of the independent variable are spaced equally far apart and when all cell sizes are equal. These conditions would normally be satisfied in research aimed at evaluating trend, because the form of the data should be evaluated with equal emphasis over its entire range.

Testing for trend involves decomposing the sum of squares for conditions. Let us suppose a researcher has run five groups and hypothesizes a linear trend over the five means. The statistical evaluation of this hypothesis of linearity involves two steps. First, there must be significant linear trend. Then it must also be shown that the remaining sum of squares, that is, the deviations from linearity, is not significant. The linear component would first be tested with 1 *df*. Then the deviations, on 3 *df*, would be tested. Notice that there is no ordinary test of the conditions sum of squares in this procedure. All of the *df* generated by the five groups are used up in the evaluation of the linearity hypothesis. In the same way, one could test for quadratic trend in the data, first by testing the quadratic component for significance and then by testing the remainder for nonsignificance. It is important to note that a significant quadratic component does not by itself imply that the data can be described by the parabola (bowl-shaped curve) that is characteristic of a quadratic function. It is also necessary to demonstrate the nonsignificance of the remainder. A graph is usually informative, but one should avoid connecting the points as the lines tend to impose structure.

By far the most common trend hypothesis is that of linearity. This is hardly surprising, since straight-line functions arise naturally in behavioral theorizing. For example, an industrial psychologist told an insurance company that her Management Assisted Motivational Analysis (MAMA) program would boost sales; and the more time the sales personnel spent in the training sessions, the larger would be the sales volume. The 30-person sales force was randomly divided into six groups, and the groups respectively received 2, 4, 6, 8, 10, and 12 hours of MAMA training. Then sales during the next month were monitored to assess the effectiveness of the program. Here are the sales figures:

	Hours in MAMA					
	2	4	6	8	10	12
	$600	$550	$645	$1050	$1000	$1110
	$1040	$680	$530	$790	$1640	$1390
	$830	$1200	$1400	$1450	$1100	$900
	$475	$950	$1130	$350	$1270	$1850
	$360	$735	$900	$925	$1130	$1645
	$3305	$4115	$4605	$4565	$6140	$6895

The ordinary ANOVA provides the between groups effect that will be decomposed to yield the trend components.

Source	df	SS	MS
Between groups	5	1,773,737.50	(not computed)
Within groups	24	2,548,100.00	106,170.83

In order to compute the linear component of the group effect, we can either derive the coefficients, get them from a table, or let COMPARISONS do the work. Coefficients are then applied to the group totals. There are six groups, so a good set of coefficients for the linear trend is −5, −3, −1, 1, 3, 5. Notice that the coefficients are spaced with equal intervals; this spacing is characteristic of the coefficients for the linear component. The standard formula for a comparison is applied.

$$SS_{\text{Linear}} = \frac{[(-5) \cdot 3305 + (-3) \cdot 4115 + (-1) \cdot 4605 + 1 \cdot 4565 + 3 \cdot 6140 + 5 \cdot 6895]^2}{5[(-5)^2 + (-3)^2 + (-1)^2 + 1^2 + 3^2 + 5^2]}$$
$$= 1,643,657.79$$
$$SS_{\text{Remainder}} = 1,773,737.50 - 1,643,657.79 = 130,079.71$$

Now we construct the ANOVA table incorporating the trend components:

Source	df	SS	MS	F
(Between groups)	(5)	(1,773,737.50)		
Linear	1	1,643,657.79	1,643,657.79	15.48*
Remainder	4	130,079.71	32,519.93	<1
Within groups	24	2,548,100.00	106,170.83	

These data are consistent with the linearity hypothesis; the test is that the linear component is significant and the remainder is not. The graph supports this view.

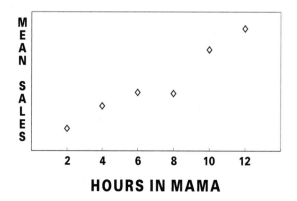

Trend tests can also be executed within a factorial design. One can test for the linear component of factor A and/or the quadratic component of factor B. The marginal totals of the factor being analyzed take the place of the group totals in our previous discussion. The appropriate error term for the trend test is whatever would have been the error term for the factor. In an independent-groups design the error is the usual within cells term.

Computer-Generated Trend Coefficients

Unequal spacing rules out use of the tabled coefficients, as do unequal cell sizes. In fact, for unequal cell sizes orthogonality may be lost so the coefficients for a component will in general not sum to zero. The algorithm for obtaining the coefficients in these cases is somewhat complex; the computer program, COMPARISONS (a successor to ORPOCO, Weiss, 1980b), can be used to find the coefficients and carry out the computations. This program, employing an algorithm previously published by Emerson (1965), requires as input the ordered values of the independent variable (referred to by the program as the abscissa values) and the cell sizes. COMPARISONS can supply fractional coefficients that have been normalized so that the denominator for the comparison is 1, or it can compute the sums of squares directly. Because of its convenience, I use COMPARISONS even in the equal spacing, equal n cases for which tabled coefficients are available. Below is the Set up Comparisons window for the MAMA problem. I set up the hypothesis of linear trend, testing the linear component and the remainder. At the moment, I am entering the abscissa value for Group 2, which is 4, that group's number of hours in MAMA.

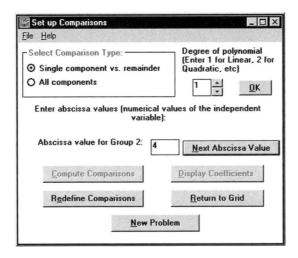

Limitations of Trend Tests

Trend tests can be disappointing in their ability to detect the true shape of the data curve. One problem is that the curve may have two components, with the second component diluting the effect of the primary one. For example, a theoretically important quadratic component may be masked by a linear component. Graphically, this situation would appear as a tilted bowl-shaped curve. Although the "bowl" aspect may be of prime theoretical interest, the tilt, which may reflect a second process or simply a scale distortion, will cause the quadratic hypothesis to be rejected. A second problem with trend tests is also related to tilting. With other things being equal, the steeper the slope of the line, the larger will be the linear component. This situation is identical to the one governing the correlation coefficient, which is also dramatically affected by the slope of the regression line; yet the steepness of the slope seems logically irrelevant to the question of whether the data follow a straight line.

Referring to the correlation coefficient calls to mind another limitation shared by trend tests. When the test confirms a hypothesis such as linearity, it is tempting to make a sweeping generalization, but it is important to remember that any conclusion applies only to the values of the independent variable actually covered in the research. Extrapolation beyond the range of the data is in general not justified.

Comparisons in Repeated-Measures Designs

Comparisons in repeated-measures designs require individualized error terms, just as tests of main effects do. Here the error terms are even more specialized, because the subject × treatment interaction for a given subset of treatments will be different from that for a different subset of treatments. Each error term will be based only on the scores involved in the particular comparison. Our goal is to obtain for any comparison both SS_{Comp} and $SS_{S \times Comp}$. As we shall see, the calculations are somewhat tedious since each subject's scores must be processed individually. COMPARISONS may be used to lighten the load.

The computational formula for SS_{Comp} is the same as that given previously for independent-groups designs. The only difference is that the n in the divisor is replaced by s; the number of subjects appears in the definition rather than the number of scores per treatment. But this difference is merely formal. In either case we divide, as usual, by the number of scores that go into the total being squared. There is one score per subject per treatment (this is true even if that score has been obtained by summing or averaging over other factors in the experiment).

$$SS_{Comp} = \frac{(\Sigma C_j t_j)^2}{s \cdot \Sigma C_j^2} \qquad (8\text{-}2)$$

To obtain the subject × comparison interaction, we weight each subject's scores by the coefficients, sum them, and then square the weighted sums. These squares are then divided by the sum of the squared coefficients. This computation generates SS_{Comp} for an individual. The individual SS_{Comp} are then summed, and the group SS_{Comp} is subtracted from that total to yield $SS_{S \times Comp}$. It is arithmetically easier to do one division rather than a separate division for each subject, and so the last summation may be carried out before the division. Here i is the index running over subjects, while the j index runs over treatments as before:

$$SS_{S \times Comp} = \frac{\sum_{i=1}^{i=s} \left(\sum_j C_j t_{ij} \right)^2}{\sum_j C_j^2} - \frac{\left(\sum_j C_j t_j \right)^2}{s \cdot \sum_j C_j^2} \qquad (8\text{-}3)$$

Let us illustrate these definitions by applying them to the numbers from the pain reliever study, with the design structurally modified to incorporate repeated measures. Thus, we have four subjects each receiving all of the drugs:

Subject	Aspirin	Blufferin	Ansin	Tyrolin	Codeine
1	31	37	27	53	10
2	23	28	39	28	15
3	42	27	26	17	13
4	39	53	29	25	22
	$t_1 = 135$	$t_2 = 145$	$t_3 = 121$	$t_4 = 123$	$t_5 = 60$

If the comparison contrasts the codeine with the other four drugs, then the coefficients are 1, 1, 1, 1, −4 and the sum of squares is given by:

$$SS_{Comp} = \frac{(1 \cdot 135 + 1 \cdot 145 + 1 \cdot 121 + 1 \cdot 123 + (-4) \cdot 60)^2}{4 \cdot [1^2 + 1^2 + 1^2 + 1^2 + (-4)^2]}$$

$$= 1,008.2$$

as before (when it was called Comp 1.)
The error term for this comparison is:

$$SS_{S \times Comp} = \frac{[1 \cdot 31 + 1 \cdot 37 + 1 \cdot 27 + 1 \cdot 53 + (-4) \cdot 10]^2 + [\cdots]^2 + [1 \cdot 39 + \cdots + (-4) \cdot 22]^2}{[1^2 + 1^2 + 1^2 + 1^2 + (-4)^2]} - S$$

$$= 1,099.6 - 1,008.2$$

$$= 91.4$$

The F ratio, on 1 and 3 df, is given by:

$$\frac{MS_{Comp}}{MS_{S \times Comp}} = \frac{1,008.2}{30.47} = 33.09*$$

These individualized error terms are computed in exactly the same way for trend tests as for other comparisons. When the scores for individual subjects are truly correlated, this analysis will show the increased sensitivity that characterizes repeated-measures tests.

The S × Remainder term needed as an error term against which to test $MS_{Remainder}$ may most easily be computed by subtraction. If $SS_{S \times Comp}$ is removed from $SS_{S \times Treatment}$, the result is $SS_{S \times Remainder}$. Alternatively, $SS_{S \times Remainder}$ may be constructed by summing $SS_{S \times Comp}$ for all of the trend components other than the desired one. For example, if there are 5 treatments, and the test is on the quadratic trend, then $SS_{S \times Remainder} = SS_{S \times Linear} + SS_{S \times Cubic} + SS_{S \times Quartic}$. This would be equivalent to $SS_{S \times Treatment} - SS_{S \times Quadratic}$.

Post-hoc Comparisons

Frequently, a researcher gets an idea after seeing the results. Effects jump out of the data, and hypotheses to accommodate them come to mind. Even more dramatically, results opposed to those originally predicted occur, and revisions in theory are called for.

Surprises must be statistically verified. And because the follow-up tests were not planned for in advance, adjustments must be made in the usual analytic procedures. The reason is that tests made afterward do not maintain their apparent significance levels. The researcher must consider the tests that might have been as well as those that were actually conducted. After all, in admitting that a particular unexpected outcome is interesting and worthy of verification, the investigator must at the same time acknowledge that some other unexpected result might equally have attracted attention.

We can highlight the problem by envisioning a six-group experiment in which the researcher's a priori hypothesis was no more specific than that there would be differences among the group means. Scrutiny of the data shows group 2 to have the largest mean and group 6 the smallest. The researcher now desires to show that groups 2 and 6 are significantly different. The obvious procedure to follow is a specific comparison, but there is a problem.

Suppose the null hypothesis is true, and all of the true group means (as opposed to the observed means obtained by sampling) are equal. The more groups, the larger the expected difference between the largest and smallest observed means. While an individual comparison is carried out at an ostensible significance level, say .05, the probability that the difference between the largest and smallest means is significant is far greater than .05. Since any of the means might have come out to be the largest and any other the smallest, we evaluate the probability of a Type I error on the basis of the number of possible pairings. The program ALPHAK may be used to find the probability. My personal view, one which is not held universally but is fairly popular among researchers, is that multiple comparison issues can be ignored when testing specific comparisons planned in advance. However, for post-hoc hypotheses, some adjustment is advisable.

Scheffé Test

Statisticians have proposed several methods to deal with the problem inherent in post-hoc comparisons. The method proposed here, due to Scheffé (1953), is perhaps the most conservative in avoiding Type I errors.[2] The Scheffé test applies a significance level to the whole set of tests taken altogether. No matter how many comparisons are made, the probability of a Type I error is limited to the specified significance level.

The Scheffé test computations for each comparison are just as they would be for any planned comparison. COMPARISONS can be used for this preliminary stage of the analysis, but the rest has to be done manually. The obtained F ratio is not compared with the ordinary tabled F ratio, however. It is compared with an adjusted F ratio. The adjusted critical F ratio for a one-way design is obtained by multiplying the tabled value for $J-1$ and $N-J$ df (where J is the number of groups and N the total number of scores) by $J-1$. That is, for the one-way design:

$$F_{\text{CRIT}} = (J-1) \cdot F_{J-1, N-J}$$

Note that the numerator df is not 1 as would be expected for a planned comparison, but instead depends on the number of groups. The test is taking into account the possibility that other group contrasts might have attracted your attention.

For a more complex design, the term $(J-1)$ should be replaced by one less than the number of totals from which the particular comparison has been selected, and the $N-J$ should be replaced by the error df for the factorial design. Comparisons can be carried out at the level of the margins or at the level of individual cells. The general expression is:

$$F_{\text{CRIT}} = (\text{Comparable totals} - 1) \cdot F_{\text{Comparable totals} - 1, \, df \, \text{Error}}$$

For example, in a 3×5 design with two scores per cell, a comparison of two of the second factor marginal totals would employ a critical F of $4 \cdot F_{4,15}$. (The 4 is there because there are five marginal totals associated with the second factor, while 15 is the error df for the factorial design. The marginal totals are the comparable totals in this case). If the comparison were to be made on a pair of cell totals, the critical F would be $14 \cdot F_{14,15}$. (The 14 is there because there are 15 cell totals that might have been compared, so there are 15 comparable totals; the error df for the factorial design remains at 15 irrespective of which comparisons you conduct.)

To get the required mean squares, first carry out a factorial analysis with FACTORIAL ANOVA. Enter the data with replicates as the first factor and with the other factors designated so that order of entry matches the order that will be used when the groups are entered in the COMPARISONS program. In that way, the scores can be copied to the Clipboard via the Data Editing menu in FACTORIAL ANOVA then pasted into the Data Entry grid via the Data Editing menu in COMPARISONS. From the factorial table, note the error df and mean square. When you exit FACTORIAL ANOVA, the scores will remain on the Clipboard. Tell the COMPARISONS program that the number of groups is the number of comparable totals. The number of scores per group will be equal to the number of replicates only if you are comparing at the level of individual cells; if you are comparing marginal totals, the number of scores per group will be the product of the number of levels of all factors (including replicates) except the factor whose marginal totals are being compared.

It should be apparent that the obtained F ratio required for a comparison to be significant must be much larger when the comparison is post hoc than when it is planned. The logic is that a truly sizable difference between means is required before the researcher can be confident that it is reliable, that it is not a chance outcome that has been seized upon. This conservatism may be thought of as the price of ignorance; unpredicted effects must be striking.

Generality is a major advantage of the Scheffé test. It can be used for all types of comparisons, including trend tests, and for as many comparisons as desired. The experimentwise Type I error rate is held under control, no matter how many tests are carried out. And like all comparison procedures, the Scheffé test affords specific, useful information.

Tukey Test

An alternative for coping with error inflation is the HSD (Honestly Significant Difference) procedure developed by Tukey (1953). It is limited to pairwise comparisons, but allows testing any pairs while still maintaining a specified experiment-wise significance level. Tukey's HSD procedure is built into the ONEWAY program because it is often employed after examination of the means. As implemented in ONEWAY, the program compares all possible pairs of means; the analyst looks at as many as are of interest. The test statistic for each comparison is the square root of twice its F ratio. This value is then compared to critical values from what is known as the Studentized-range distribution. Although the test is usually carried out on a post-hoc basis, it is perhaps more appropriate as a planned procedure. Requiring overall significance prior to testing particular pairs reduces the power of the entire analysis (Bernhardson, 1975).

Exercises

8-1. Three methods of teaching foreign language vocabulary were compared. To evaluate them, a 50-item test was administered to the 24 students in the experiment, 8 in each group. The data, expressed as a number of correct items out of 50 for each student, were as follows:

Method 1: 19, 37, 28, 31, 29, 20, 36, 33

Method 2: 21, 18, 15, 23, 20, 22, 26, 14

Method 3: 17, 20, 28, 30, 13, 18, 19, 23

Test the hypothesis that methods 1 and 2 differed from method 3 in their effect. How many additional comparisons can be constructed that are orthogonal to this one? If there are any, what might they be?

8-2. The inverted-U theory of motivation holds that an animal with little motivation will not work hard, nor will an animal with extreme motivation (perhaps because the beast cannot concentrate on the task). High performance will arise from an intermediate level of motivation. This theory can be translated into a quantitative prediction of a quadratic trend.

The theory was tested by an enterprising graduate student who withheld food from 24 rats for specified intervals. After the requisite starvation period, each rat was tested in a complex maze using a food reward. Once a day, each subject was returned to the maze for a single trial, after having been deprived for the designated number of hours. The rat finished his servitude when he had performed the maze in an errorless fashion. The score was the number of trials required until this criterion was achieved. Evaluate the theory with these data.

Number of Trials Required to Navigate Maze Perfectly

Hours of deprivation			
1	7	13	19
15	8	11	21
11	10	7	12
14	7	11	19
9	5	13	14
16	9	10	20
14	8	11	16

8-3. The following data are correct responses, out of a possible 25, on four difficult concept learning tasks, involving a total of 20, 40, 60, and 80 trials, respectively.

	Total trials			
Subject	20	40	60	80
S_1	11	10	15	23
S_2	2	5	9	13
S_3	6	3	12	13
S_4	7	4	13	15
S_5	11	12	9	19
S_6	7	5	12	20

Test the hypothesis that number correct is a linear function of total trials.

8-4. As an innovative professor, I conducted an experiment with independent-study students. I was interested in getting the students to run lots of subjects in their projects. I conducted a factorial study of instruction in which I crossed the factors of threats and promises. The four levels of threats were: (1) threaten bad grade; (2) threaten to drop from independent study; (3) threaten to cut off financial support; (4) threaten to make student be a subject in an electric shock experiment.

116 ANALYSIS OF VARIANCE AND FUNCTIONAL MEASUREMENT

The three levels of promises were: (1) promise good letter of recommendation; (2) promise coauthorship on paper; (3) promise to send student to report results at convention in Paris. Each of 60 students was given one threat to be delivered if performance was "bad" and one promise to be kept if performance was "good." The dependent variable of interest was the number of subjects run by each student. The results of this study were as follows:

Threat$_1$			Threat$_2$			Threat$_3$			Threat$_4$		
Promise$_1$	Promise$_2$	Promise$_3$	Promise$_1$	Promise$_2$	Promise$_3$	Promise$_1$	Promise$_2$	Promise$_3$	Promise$_1$	Promise$_2$	Promise$_3$
8	4	5	2	2	3	3	3	3	14	18	14
5	4	6	4	6	4	8	3	6	12	14	13
2	4	2	5	9	8	2	5	6	9	11	3
3	3	6	7	4	9	2	6	7	15	9	12
5	2	2	3	5	7	4	7	5	18	11	11

Perform the analysis of variance of these data. Scanning the data suggested that the electric shock threat produced very large responses. Perform the appropriate test to confirm the inspection of the data. (Hint: The COMPARISONS program only handles one-way designs, so you will have to lie to it.)

8-5. Each volunteer was given a dose of a painkiller whose side effects were being evaluated, and after 20 minutes each volunteer was given the same standard anagram to solve. The solution times, in minutes, are given below.

		Drug dosage				
		Control(0)	1 gram	2 grams	3 grams	7 grams
Males		4	10	5	9	15
		4	14	4	15	8
		9	14	9	11	15
		7	18	5	13	10
Females		12	9	11	14	15
		2	12	14	8	2
		1	8	15	17	7
		10	3	16	17	8

(Repeated hint: Because COMPARISONS only handles one-way designs, you'll have to be creative to get the appropriate error term.)

(a) Test the hypothesis that there is no difference between the overall mean for the control group and the average of the four experimental group means.

(b) Test the hypothesis that there are no overall differences among the four experimental group means.

8-6. A graduate student conducted a research project on the effect of coin incentives. Nine-year-old children were individually asked to memorize a list of twenty words. The number of words correctly repeated after a 2-minute waiting period was the dependent variable. Each child was promised either a penny, a nickel, a dime, a quarter, or a half-dollar per correct response. Children were randomly assigned to a particular reward condition. The hypothesis was that the number of words recalled would be a linear function of the value of the incentive. Unfortunately, the graduate student ran out of money before all of the data had been collected, and so the planned equality of group size was not achieved. Evaluate the hypothesis of linear trend.

Penny	Nickel	Dime	Quarter	Half-dollar
7	9	8	5	18
3	13	4	15	10
5	6	10	12	14
8	5	14	9	13
4	12	7	16	
	5		11	

8-7. Cursory examination of the results of exercise 2-5 suggests that pies in faces inspire more laughter than snappy insults. Use the Tukey HSD test to confirm the suggestion.

Answers to Exercises

8-1.
Source	df	SS	MS	F
Comparison	1	65.33	65.33	2.09
Within groups	21	657.75	31.32	

There is one additional orthogonal comparison possible; it tests the hypothesis that the impacts of methods 1 and 2 differ.

8-2.

Source	df	SS	MS	F
(Deprivation)	(3)	(275.46)		
Quadratic	1	210.04	210.04	31.55*
Remainder	2	65.42	32.71	4.91*
Within groups	20	133.17	6.66	

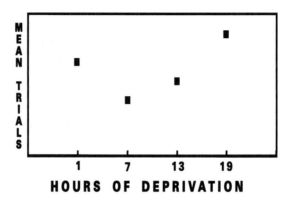

HOURS OF DEPRIVATION

Note that the quadratic hypothesis is not supported, because the remainder is significant.

8-3.

Source	df	SS	MS	F
Subjects	5	153.33		
(Trials)	(3)	(430.33)		
Linear	1	360.53	360.53	93.56*
S×L (error)	5	19.27	3.85	
Remainder	2	69.80	34.90	5.59*
S×R (error)	10	62.40	6.24	

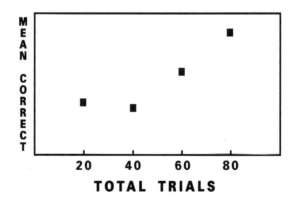

TOTAL TRIALS

8-4.

Source	df	SS	MS	F
Threats	3	663.25	221.08	32.35*
Promises	2	0.10	0.05	<1
TP	6	43.50	7.25	1.06
Within cells	48	328.00	6.83	

Post-hoc comparison:

$$\frac{F_{obs}}{6.83} = 653.61 = 95.70 *$$

$$F_{crit} = 3 \cdot F_{3,48} = 8.40$$

8-5(a).

Source	df	SS	MS	F
Comparison	1	150.16	150.16	10.94*
Within cells	30	411.50	13.72	

The comparison is carried out as though the design were a one-way with five groups. However, note that the gender main effect and interaction have been removed to purify the within-cells error term. A factorial ANOVA preceded the comparison to accomplish the removal.

(b) This should be done as a standard, two-way ANOVA on the four groups selected.

Source	df	SS	MS	F
Gender	1	0.03	0.03	<1
Treatments	3	50.09	16.70	1.33
GT	3	248.09	82.69	6.60*
Within cells	24	300.75	12.53	

8-6.

Source	df	SS	MS	F
Linear	1	167.14	167.14	14.05*
Remainder	3	20.93	6.98	<1
Within groups	21	249.82	11.90	

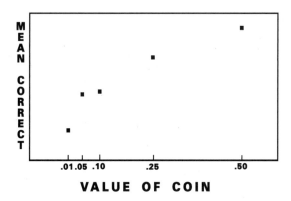

The hypothesis is supported (linear component is significant, remainder is not).

8-7. The test statistic for the comparison, Sqrt($2F$), is 4.70, which is significant at the .05 level. So there is a significant difference between pies and insults.

Notes

1. Gaito (1978) has argued for the practical value of nonorthogonal comparisons in research. He contends that answering questions of empirical concern has top priority. The maintenance of independence is, in his view, less important than the meaningfulness of the comparisons.

2. Accordingly, the Scheffé test is weak in terms of power. The researcher must decide whether Type I errors or Type II errors are the greater danger. A comprehensive discussion of alternative multiple comparison procedures is given in Toothaker (1991).

9

Measurement Issues

Although practitioners often regard measurement concerns as a side issue, there is a sense in which measurement is the key problem in behavioral research. The researcher's goal is to know and to understand how variables affect one another. We must be able to control changes in behavior. The assessment of change requires measurement.

Statistical tests are distinct from measurement questions. We can compare sets of numbers that have no behavioral referents. But for the researcher, such comparisons are pointless exercises; a statistical test is meaningful only to the extent that the data being processed are meaningful. A statistical test answers questions about numbers, but the researcher's interest is in what the statistical tests tell us about the behavior they measure.

This distinction can be clarified with a facetious example suggested by Lord (1953). Suppose I want to make a comparison of two football teams—let's say Cal State and Notre Dame. Since the teams have not played each other recently, I decide to compare the players themselves. As my numerical measure, I use the convenient numbers placed on the players' backs. I carefully perform a one-way ANOVA on the two sets of scores, and find them significantly different, with Cal State having the higher mean. So I proudly announce that statistical analysis has scientifically shown Cal State to be the better team.

Before a challenge is issued and I have dead Cal State players on my conscience, let me hasten to admit that my conclusion is nonsense. But what is the nature of my error? Is it that one cannot perform ANOVA on football numbers? For those readers who are not sports fans, let me point out that football numbers are simply identification labels; the technical term is that the numbers are measures on a nominal scale. Stevens (1951) has argued that proscription is in order. He points out that the computations of ANOVA involve arithmetic operations—summing, multiplying, and so on, and that such operations are peculiar things to do to labels. Stevens deems the operations invalid, and thus any analysis employing them is invalid as well.

But that's not my error; Stevens has missed the point. Any set of numbers is grist for the ANOVA mill. The mathematical machinery does not depend upon the origin of the numbers processed. I can compare the two sets of football numbers and find them different. What I cannot do is thereupon to make any inference about football. My statistical test told me only that one set of football numbers is larger than the other; it answers the question "Are the means of the two sets of football numbers different?" This question is of no interest, because football numbers do not measure any underlying property of football behavior, but ANOVA does indeed give a correct answer to it. My error is to make an inference about football from a set of scores unrelated to football ability. Football numbers are invalid as a measure of any important property; therefore, a conclusion that depends upon these numbers is invalid as well.

Whether a measure is valid depends upon its relation to the behavioral property of interest. Justifying the use of a particular measure is one of the researcher's important responsibilities. Traditionally, "face validity" (does the measure seem reasonable?) has been the primary criterion, but reasonableness is sometimes in the eye of the beholder. Few would argue against counting hours of deprivation as a measure of the strength of a rat's hunger, but what about counting hours spent in church as a measure of the strength of a person's religious feeling? A statistical inference is valid only to the extent that the measures it is based on are valid.

Controversy over the connection between scale properties and permissible statistical procedures has continued (indeed, even raged on) for years. The "permissive" position, that ordinal properties are sufficient to justify ANOVA, has been defended by Anderson (1961) and by Gaito (1980). The "restrictive" perspective, that interval data are necessary for ANOVA, has been advanced by Suppes and Zinnes (1963) and by Townsend and Ashby (1984). The restrictive position is that the calculation of variances (and even means) implies that the difference between numbers is independent of their location on the scale. A measured difference of two points on the scale, such as the difference between 2 and 4, has to reflect the same psychological difference as that between 32 and 34 or between 92 and 94. Interval scales and ratio scales satisfy the requirement of equality of intervals, but ordinal scales do not.

The problem for behavioral researchers is that plausible ordinal measures are relatively easy to construct, but convincing arguments for the interval relationship between a behavioral index and the underlying property it purports to measure are hard to achieve. I am willing to accept that when a respondent reports intensity of feeling, a larger number truly implies a more intense feeling. However, when someone reports liking Merlot "6" and Chardonnay "3" on a 10-point scale, I am skeptical that the responses are quantitatively meaningful.

The permissive folks contend that in practice, it seldom matters whether the equality of intervals requirement is satisfied. My own foray into this minefield (Weiss, 1986) was the empirical demonstration that F ratios for interval data of unquestioned validity did not change significantly when the scores were artificially degraded so as to have only ordinal properties. A possible explanation of this result is given in a cogent delineation of the conditions under which ordinal data will yield valid statistical inferences (Davison and Sharma, 1988). The empirical result, along with the pragmatic conviction that much useful research would grind to a halt if we were forced to produce interval measures, leaves me firmly embedded in the permissive camp.

Technical Assumptions

Like any statistical technique, ANOVA applies in a specific set of conditions, conditions that in principle must obtain if the analysis is to be correct. The cautious phrase *in principle* embedded in the previous sentence is intended to imply that violations of these assumptions are not necessarily fatal to the analysis.

The first class of assumptions to consider includes two distributional notions. The NORMALITY assumption means that each score is presumed to be drawn from a normal distribution. This assumption can be checked, if the group contains enough observations, by comparing the obtained scores against a theoretical normal distribution using a χ^2 test. The HOMOGENEITY OF VARIANCE assumption means that all of the groups have approximately the same variance. Homogeneity can be assessed with Bartlett's F max test.

In current practice, researchers pay little attention to these distributional assumptions: they are rarely checked. The justification for this seeming laxity is that simulations with artificial data have shown even severe violations to have little effect on significance statements produced by the analysis. A simulation is conducted (usually on a computer) by constructing many sets of data in each of which the group means are equal (so that the null hypothesis is true) but a distributional assumption—for example, normality—is violated. If the ANOVAs conducted on these artificial data sets reject the null hypothesis in approximately 5% of the cases, then one may conclude that the significance test is not seriously affected by the violation. ANOVA has been shown via such simulations to be robust, that is, virtually unaffected by even severe nonnormality or by gross heterogeneity (nonhomogeneity) when the group sizes are moderately large.[1]

Another technical assumption underlying some ANOVAs is additivity of effects. This phrase refers to the idea that certain sources must not interact. For example, in repeated-measures designs, each source is tested against its interaction with subjects. Such a test makes sense only when the true interaction between the source and subjects is zero, and thus the mean square for the interaction contains only the random error component (e_i). To the extent the true interaction is nonzero, the significance test is insensitive. In the confounded designs we shall discuss in chapter 13, additivity is even more crucial. Designs are used that, for want of anything better, employ interaction between substantive sources as error terms. If the interaction-used-as-error is not truly zero, the entire analysis may be invalid.

Additivity of effects is a crucial assumption in many analyses, and it cannot be dismissed on the grounds of ANOVA's robustness as researchers do with the other technical assumptions. It is checked upon less often than it should be. An interaction that is truly zero should be nonsignificant in practice; to assess the significance of the interaction, a within-cells error term against which to test it is required. For a repeated-measures design, the conceptually simplest way to get such an error term is to replicate the experiment. Since replication is sometimes infeasible, an alternative approach is to conduct a pilot project prior to the intended research. In this pilot project, a small-scale within-groups design may shed light on the magnitude of the critical interaction terms. If these terms prove small enough, the researcher may have some confidence that nonadditivity of effects will not bias the important conclusions.

Independence among the scores is a different kind of assumption, one required by experimental logic. In an independent groups design, it is obviously important to keep the participants from influencing one another's responses. Less obvious, perhaps, is the need to make sure that the equipment or response instrument does not constrain the responses so that people will respond similarly even though their responses would vary under optimal conditions. In a repeated measures design, of course, scores from the same person cannot be regarded as independent. However, those scores should be as free from carry-over effects as possible. In particular, the researcher should try to inhibit respondents from using remembered previous responses as a basis for later responses. No statistical magic can undo the damage caused by violations of independence, because some of the scores that get analyzed are "wrong."

The Heartbreak of Zero Error Variance

Occasionally an error mean square will be zero. This presents a problem for your calculator or computer at F ratio time, since dividing by zero is an insurmountable challenge. While in principle it is possible for the true value of the error variance to be zero, in practice the value should usually be rejected. Experimentally, zero variance probably means either that the response instrument is too crude to detect real differences or that the research technique is inducing dependent responses (e.g., the subject remembers and is trying to be consistent with earlier responses). It is likely that the data are invalid, and a return to the laboratory is in order.

Transformation

What happens when the data do not meet the technical requirements of ANOVA? During the 1940s, an ingenious method of salvaging a set of data was introduced, one that has remained controversial to this day. A TRANSFORMATION is a mathematical function that gets applied to all of the scores; in general, the transformation may be written as $T_i = f(x_i)$, where f is a specified rule. Transformations often take the form of a simple operation, such as the square root ($T_i = \sqrt{x_i}$), the logarithm ($T_i = \log x_i$), or the arc cosine ($T_i = \cos^{-1} x_i$). Sometimes a discontinuous function is used ($T_i = x_i + 3$ if $x_i < 10$; $T_i = \max(x_i, 7)$ if $x_i \geq 10$). Any transformation that is not linear (that is, that cannot be written as $T_i = ax_i + b$, where a and b are constants) will affect the ANOVA F ratios. A linear transformation leaves all F ratios unchanged.

A transformation is applied in the hope it will eliminate (or at least reduce) the problem in the data set. Typically, the early transformers applied their magic on an ad-hoc basis; the criterion for the choice of transformation was whether it worked. For the popular goal of variance stabilization, the square root transformation was employed when the variance for each group was correlated with the group mean. The arc sine transformation ($T_i = \sin^{-1} x_i$, commonly referred to as arcsin) was often employed by learning theorists because it attenuated the "ceiling" effects that arise with results expressed in percentages. More systematic transformation approaches were developed later; one very sensible plan was to use a power transformation ($T_i = ax_i^p$), selecting the value of the exponent (p) that best achieved the particular goal. Computer programs have been developed that put an extremely powerful weapon in the transformer's armory. Kruskal's (1965) MONANOVA program adjusts the data until nonadditivity has been made as small as possible given the restriction that the transformation must preserve the ordering of the scores. FUNPOT (Weiss, 1973) and POLYCON (Young, 1972) are similar but more general rescaling routines.

Researchers who used transformations were sometimes pleasantly surprised to find that a transformation employed to achieve one technically desirable goal often accomplished as a fringe benefit other statistically worthwhile ends. For example, a transformation designed to stabilize variance might well simultaneously reduce nonnormality and nonadditivity of effects. Believing these concomitant benefits could hardly be accidental, some investigators developed the rather mystical notion of a proper metric. The point was that when a behavioral variable was measured on the truly appropriate scale, the scores would automatically satisfy the statistical criteria. Thus, statistical optimality could be used as a validating criterion for a scale, providing a criterion that transcends face validity.

Transformation functions can be applied using the Transformation button available in the Data Entry window of all of the CALSTAT programs that process numerical data. The selected function is applied to all of the scores in the data set. The choices available include linear transformations (adding a constant or multiplying by a constant), which change values but not F ratios, and four nonlinear transformations, which do change F ratios. The rank-order transformation replaces scores with their ranks within the data set. In the example below, a square-root transformation has been set up. The Revert to OK'd Data button undoes all transformations and restores the original data.

126   ANALYSIS OF VARIANCE AND FUNCTIONAL MEASUREMENT

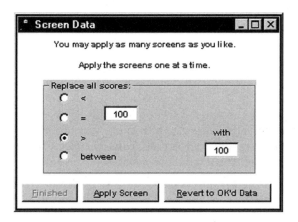

Another kind of transformation is offered via the Screen Data button, also available in the Data Entry window. These options allow the user to replace all scores in a designated region with a particular value. In the illustration below, any score entered as greater than 100 (the defined maximum on the response scale) was considered to be a mistake, and was replaced with 100. The primary intent underlying Screen Data is to cope with errors committed by the respondents; but the option can also be used to impose scale adjustments on the responses, such as a response ceiling suggested by theory.

Most researchers have been very conservative with respect to transformations. Tampering with the data seems dangerously close to dishonesty, and post-hoc analyses seem to give transformation users several turns at bat whereas the traditional researcher has only one. The counter to the conservative position is that there is nothing sacred about any particular measuring scale. For example, we conventionally measure length with an inch ruler or a centimeter ruler; since the two scales are related by a linear transformation, they are equivalent for ANOVA purposes. But now, imagine a society that measures length with what we would call a slide rule; the graduations on their rulers would be logarithmically related to those on ours. Operations on lengths in that society would be orderly and internally consistent, but they would not be the familiar ones. For example, if I measure two sticks and then lay them end to end, I find the length of the combination to be numerically equal to the sum of the individual lengths. This property of length is additivity, a term we have seen in other contexts. Additivity does not hold for slide rule length. In the slide rule culture, the length of the combination is equal to the product of the two lengths (check this for yourself). Children would learn the law: length (A + B) = length (A) · length (B); length would be multiplicative.

The slide rule example illustrates the idea that one measuring scale cannot be logically superior to an ordinally equivalent competitor. Our choice of scales is in this sense arbitrary, and we might as well base the selection on statistical convenience as on familiarity. The fact that scores in maze-running experiments are usually collected as speeds (feet per second) need not constrain us to carry out the ANOVA using speed scores. If transforming the scores into durations (seconds spent in the maze) makes them more tractable to analysis and this isdeemed important, then one should not be intimidated by charges of impropriety.

The neutral position toward measurement scales I have just expressed is certainly arguable. The fact that a measurement scale yields statistically comfortable data does not necessarily mean the measures convey accurately the underlying behavioral property. I lean toward the functional measurement (Anderson, 1976) view of validity. A scale is validated when data gathered with it confirm a behavioral law; measurement and theoretical understanding go hand in hand. This perspective will be developed in chapter 14.

Functional measurement also allows for the use of transformation, but the goal of these adjustments is not statistical in nature. Transformation may be employed to bring the data in line with the behavioral theory. Such procedures seem dangerous, since one might be able to support any theory. However, only monotone (order-preserving) transformations are permitted. The idea is that one would not want to reject a model of behavior simply because an inappropriate measuring device was used; the rather spare assumption that responses have the correct order is preferred. Empirically, models have been rejected (Anderson, 1977). The key safeguard against indiscriminate transformation is the requirement that scales measuring similar objects in different experimental settings must converge.

Exercises

9-1. Is there anything important between the two ears? A psychoacoustician explored this question by presenting tones of varying intensity to each ear simultaneously. The highly trained single respondent's task was to estimate the total loudness of each presentation. Responses were made using the method of magnitude estimation, which Stevens's theory suggests should be logarithmically transformed prior to analysis. Three replications were carried out. Analyze the (transformed) responses. If you are not a Stevens disciple, you might wish to analyze the raw data as well.

		Left ear intensity (dB re SPL)			
		10	20	30	40
Right ear intensity (dB re SPL)	10	6, 5, 9	27, 20, 22	25, 20, 16	22, 29, 31
	20	17, 19, 22	20, 24, 25	32, 25, 28	33, 37, 39
	30	28, 30, 25	35, 40, 36	46, 39, 44	51, 46, 43
	40	36, 30, 36	42, 50, 38	51, 43, 42	60, 45, 39

Answers to Exercises

9-1.	Source	df	SS	MS	F
	Right	3	6.21	2.069	89.92*
	Left	3	3.54	1.179	51.26*
	RL	9	1.60	0.178	7.72*
	Within cells	32	0.74	0.023	

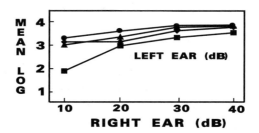

Note

1. For factorial designs with unequal cell sizes, this casual reliance upon robustness will not do. See chapter 12, note 1, for further information.

10

Strength of Effect

Despite the acknowledged usefulness of ANOVA in settling the central question of whether an experimental manipulation affects the response, some researchers have expressed dissatisfaction with an inherent limitation of the technique. The limitation is that the machinery for testing a hypothesis does not disclose the magnitude of the manipulation's effect. An effect may be significant, that is, reliable, but at the same time it may be small. And small effects do not provide the satisfaction that comes from demonstrating clear-cut experimental control.

It is tempting to use the significance level of an F ratio as an index of the magnitude of the effect. Significance at the .01 level is seen as more impressive than significance at the .05 level. This temptation should be avoided, for its underlying logic is incorrect. The problem is that significance depends not only on the size of the effect but also on the number of scores. Indeed, virtually any small true difference can be made a significant one if the researcher gathers enough observations.

This point is not well understood, even among experienced scientists. Rosenthal and Gaito (1963) asked 19 faculty members and graduate students, all of whom had conducted research projects, for ratings of their degree of belief in hypothetical research findings. The contrasts of interest were between experiments whose results reached a given level of significance based on cell sizes of either 10 or 100. For every significance level, the researchers expressed more confidence in the experiment based on the larger number of observations. From an objective point of view, the researchers were wrong. To achieve a given level of significance with a smaller number of scores, the effect must be stronger. Thus, more confidence should be placed in the results based on 10 observations, given that the same level of significance was attained in both instances.

Of course, this argument does not imply that it is better to conduct experiments with small N; the inherent lack of power with few observations may cause the researcher to fail to confirm real effects. Indeed, Maxwell (2004) has argued cogently that underpowered studies are a primary reason for the lack of cumulative progress in psychology. The assertion is simply that a significant result obtained with small N is a more impressive demonstration than a comparable one obtained with large N. A researcher may elect to gamble by deliberately choosing to collect as few observations as is felt will be needed to achieve the hypothesized result; the reasoning behind such a strategy is that significance achieved with a large number of observations may be viewed by the scientific community as unimportant.

To estimate the required number of observations requires an assessment of the statistical power in the experiment. Power depends upon many aspects of the experimental setting, including procedure and method of analysis as well as the magnitude of the effect under study. The determinant that is easiest to regulate is the number of observations gathered. In this context, the power tables given by Cohen (1968) can give useful information about the optimal number for a given experiment. Cohen's work is much admired but does not always affect research practice. The difficulty, in my view, is that calculating power requires advance knowledge of the effect size. Unless the research is closely modeled on a previous study, this information is seldom available. Experimenters rely on intuition and hope that pilot results hold true.

Instead of looking only at significance, some researchers have proposed direct measures of the magnitude of an experimental effect. The goal is to assess the importance of the results in a way that does not depend upon the number of observations. The fifth edition of the *Publication Manual of the American Psychological Association* urges authors to report effect size measures; some journals even require them. A justification of this policy is put forth in Wilkinson and APA Task Force on Statistical Inference (1999). I believe that compulsion goes too far, as in some cases effect size measures can be misleading, and I would prefer to leave the decision to the researcher. Nevertheless, the policy confirms the belief of many scientists that standard significance tests fail to capture an important aspect of the data. Reporting **confidence intervals** in addition to means can convey useful information.

Hays's ω^2

A widely accepted solution to the problem of measuring the strength of an effect was proposed by Hays (1963), and the measure is known as ω^2 (omega-squared). Hays showed that information already contained in the ANOVA table can be used to assess magnitude of effect and argued that merely conducting significance tests was using the data inefficiently. ω^2 is a measure of what Hays called statistical association. Association refers to the connection between an experimental effect and the response; it is much like correlation. It may be viewed as the proportion of variance accounted for by the effect. ω^2 contributes an indirect measure of the power of the associated F test (Keren & Lewis, 1979).

For fixed effect designs with equal cell n, the computations for ω^2 are simple.[1] Recall that fixed effect designs are those without repeated measures, designs in which the within-cell mean square serves as the error term. The estimation formula is equation 10-1:

$$\omega_A^2 = \frac{SS_A - (df_A \cdot MS_W)}{SS_{Total} + MS_W} \qquad (10\text{-}1)$$

Here A is any source; MS_W is the mean square within, and $SS_{Total} = \Sigma X^2 - T^2/N$. Computation of ω^2 can be illustrated with the data from chapter 5's example, the mathematics learning study. Let us first examine the highly significant Methods effect, for which the F ratio was 12.43. Using equation 10-1, we obtain

$$\omega_{Methods}^2 = \frac{7100.72 - (2 \cdot 285.57)}{16427.56 + 285.57} = 0.39$$

Thus almost 40% of the variance in the scores is accounted for by the differences among the teaching methods. In contrast, it should be intuitively clear that the nonsignificant Calculators effect will yield a small ω^2. Plugging the numbers into equation 10-1, we obtain

$$\omega_{Calculator}^2 = \frac{.444 - (1 \cdot 285.57)}{16427.56 + 285.57} = -0.017$$

The negative value obtained for a quantity that purports to be a proportion should be mildly disturbing. The resolution is that ω^2 is an estimate, and is thereby subject to sampling error. In fact, the same random occurrences that can result in an F ratio being less than one can also result in a negative estimate of ω^2. The negative value should be interpreted as if it were zero. However, the actual value should be given in a research report.

There are no rigid guidelines for interpreting obtained values of ω^2; but large values, and more importantly small values, speak for themselves. Although they are less familiar to researchers than are significance statements, the percentage-like ω^2 values convey information that is intuitively clear. The ONEWAY program offers an ω^2 button in the window that displays the Anova table, as does FACTORIAL ANOVA; click the Hays, P-h P option in the latter case.

Evaluation of ω^2

Most statistics texts consider ω^2 a praiseworthy statistic and recommend it be used routinely. However, few researchers have incorporated it into their repertoire. Probably the reason for this reluctance is that ω^2 is easy to use in a misleading way. A researcher can regulate the percentage of variance accounted for by a source by adjusting its levels. The farther apart are the levels, the greater will be ω^2. Because the choice of levels is usually under control of the experimenter (Glass & Hakstian, 1969), the skeptical research community is reluctant to place too much faith in the reported degree of statistical association. Of course, the choice of levels similarly determines the magnitude of the F ratio. But while consumers of research who can appreciate significance statements inevitably have some sophistication in these matters, even the untutored are likely to think they understand a statistic that looks like a percentage. It is only for experimental factors whose levels have naturally occurring values (such as gender or ethnicity) that ω^2 can convey without reservations a feeling for the importance of the effect.

Even with such factors, a multifactor design presents difficulties. The proportion of variation attributed to one factor can be affected by the spacing of the levels on other factors (Maxwell, Camp, & Arvey, 1981). While the magnitude of any effect is obviously dependent on the values chosen by the researcher, the comparative nature of the ω^2 statistic makes it especially difficult for the scientific audience to keep the limitation in view. When one is told that a given variable determines much more of the variance than another in the same experiment, it is natural to conclude that the larger proportion corresponds to a more powerful factor. A very cogent statement of the problems inherent in the use of ω^2 has been given by O'Grady (1982). In studies in which it is of practical importance to compare proportion of variance across factors, perhaps a way to mitigate the problem is to use the most extreme naturally occurring levels on all variables. In that way, the charge of arbitrary selection of levels in order to promote a favored factor can be refuted.

A modification of ω^2 suggested by Keren and Lewis (1979) may offer promise. They argued that a partial ω^2, so named because it is analogous to a partial correlation, would in general be more informative than a standard ω^2. Effects other than the one under consideration are partialed out, which alleviates the criticism that one effect's proportion of variance inevitably diminishes the proportion attributed to other effects. With the partial ω^2, the impact of each substantive source is assessed independently of the contribution of the others. For an independent groups design, the estimation formula is equation 10-2:

$$\text{partial}\,\omega_A^2 = \frac{SS_A - (df_A \cdot MS_W)}{SS_A + (N - df_A) \cdot MS_W} \qquad (10\text{-}2)$$

One may apply equation 10-2 to each source of interest, including interactions. Each estimate conveys the extent to which the source affected the response. In evaluating the effects obtained in an experiment, one must still consider that the researcher's choice of levels can determine the partial ω^2, but one need not fear that the choices for one variable have inflated or deflated the apparent magnitude of effect for another. While the partialing does seem to afford a better mousetrap than the standard ω^2, researchers have not beaten a path to its door. Perhaps the reason for this apathetic reaction is that the simple interpretation as proportion of variance is not available with the modification (because the partial

ω^2 values do not add up to one), although in my view that is an advantage rather than a shortcoming.

The data from the example given in chapter 6 serve to illustrate partial ω^2. For the source denoted A, equation 10-2 is filled in to yield:

$$\text{partial } \omega_A^2 = \frac{173.06 - (3 \cdot 4.73)}{173.06 + (48 - 3) \cdot 4.73} = 0.41$$

Partial ω^2 values for the ANOVA table may be shown as follows:

Source	df	SS	MS	F	partial ω^2
A	3	173.06	57.69	12.20*	.41
B	1	0.19	0.19	<1	−.02
C	2	88.29	44.15	9.34*	.26
AB	3	6.23	2.08	<1	−.04
AC	6	17.88	2.98	<1	−.05
BC	2	68.63	34.32	7.26*	.21
ABC	6	39.23	6.54	1.38	.05
Within cells	24	113.47	4.73		

As is the case with ordinary ω^2, a negative value should be viewed as if it were zero. The ω^2 button in FACTORIAL ANOVA also provides access to partial ω^2; click the Partial option.

η^2

Another popular measure of effect size is η^2 (eta-squared), which was recommended by Fisher (1935). Like ω^2, η^2 attempts to convey the proportion of variance associated with a source. Its definition looks simpler than that of ω^2:

$$\eta_A^2 = \frac{SS_A}{SS_{\text{Total}}}$$

However, η^2 is a overestimate of effect size, because it estimates for the sample rather than for the population as ω^2 does. Because of the bias, I suggest using ω^2 even though the computations are slightly more complex. If the design has more than one factor, η^2 is also subject to the same dependence of one effect's size on that of the others that we saw with ω^2. Accordingly, a partial η^2 statistic has been developed. Similar to what we saw with partial ω^2, the sum of the partial η^2s for a design need not add up to one. The definition calls for a change in the denominator:

$$\text{partital } \eta_A^2 = \frac{SS_A}{SS_A + SS_{\text{Error}}}$$

Partial η^2 is subject to the same upward bias as η^2, so I recommend using partial ω^2 instead if a measure of effect size is called for in a multifactor design.

The Schumann and Bradley Test

A different approach to assessing strength of effect has been proposed by Schumann and Bradley (1959). They developed a rigorous way to compare sets of data from two similar experiments, one that results in a significance statement proclaiming one of the experiments to have produced a stronger effect. The key quantity in the formulation is the ratio of the corresponding F ratios from the two experiments. This ratio is compared to a tabled critical value in the same manner as are other obtained statistics.

The Schumann and Bradley test can be used only in a specific circumstance. The experiments to be compared must be virtually identical in structure. They need to have the same number of factors, levels, and observations per cell. For the test to be scientifically interesting, there must be a strong substantive connection between the experiments. Usually the experiments will be components of a single investigation, for only in that circumstance is the requisite control of structure likely to be achieved.

Perhaps the major importance of the test is that it can provide definitive answers about methodological questions. For example, opinion surveyors need to know the optimal number of response categories for eliciting information from respondents. Various criteria for optimality have been suggested. Following Cochran (1943), it is proposed that a better response system is more sensitive to reliable differences of opinion. Suppose two response systems are under consideration, for example, a 6-point scale and a 20-point scale. The 6-point scale may not afford a sufficiently fine grain to capture subtle opinion shades; on the other hand, the 20-point scale may give rise to excessive random variation.

For the comparison of the response systems, each scale defines a separate experiment. The proposed optimality criterion implies that if each of several respondents answers each of several questions using the two response systems, then the superior system is the one that produces the greater consistent differences in responses relative to the variability. Within each of the separate experiments the questions constitute the levels of one factor and the respondents the levels of another. The F ratio from that design is a measure of the efficacy of the response system.

In our example, the two experiments are identical in structure. The same questions are used in both, and the same number of responses is collected. The Schumann and Bradley test can be used to determine whether the scale hypothesized to be superior yields a significantly larger F ratio.

The observed statistic in the Schumann and Bradley procedure is based entirely on values included in the ordinary F table. We can illustrate the computations using the data from a study that employed the logic described above. The study was a psychophysical one, in which observers judged the average darkness of 25 pairs of gray chips using two response modes, graphic ratings and magnitude estimation (Weiss, 1972). Analysis of variance on the judgments showed that the F ratio for stimulus pairs, tested against its interaction with subjects, was 72.07 (df = 24, 168) for graphic ratings; and the corresponding F ratio for magnitude estimation was 32.30 (df = 24, 168). Both of these F ratios were highly significant; this merely means that the various stimulus pairs were judged differently. Here is the opening window using the SCHUBRAD program:

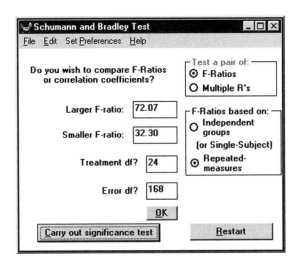

Let us denote the graphic ratings as Experiment 1, and the magnitude estimates as Experiment 2. The key quantity to be computed is W = F_1/F_2 = 72.07/32.30 = 2.23. W *is the quantity that is tested against the tabled quantity,* W_0. *The parameters a and b, based on degrees of freedom from the F table, are needed in order to enter the W table to find the appropriate* W_0; a = $df_{treatment}/2$ *and* b = $df_{error}/2$.

The triangular Schumann and Bradley table (in Schumann and Bradley, 1959) is somewhat difficult to read. Therefore, the algorithm and the table have been incorporated into SCHUBRAD, a computer program that carries out the analysis (the original version was presented in Weiss, 1985b). Inputs to the program are the two F ratios, the dfs for the two experiments, and whether the designs are independent-groups or (as in the present example) repeated-measures.

The larger F ratio for graphic ratings was expected on theoretical grounds, and so a one-tailed test is appropriate. The one-tailed test is executed by comparing the obtained value of W, 2.23, with the critical value, W_0, (from SCHUBRAD) of 2.09. As this value of W is significant at the .05 level, we can pronounce the graphic rating scale as more sensitive that the magnitude estimation technique. Had no directional prediction been made, a two-tailed test would have been in order. The latter test looks in two directions; if W is greater than W_0 or if 1/W is greater than W_0, the null hypothesis of no difference in sensitivity may be rejected.

For an independent groups design, some additional steps are needed. The quantity λ is calculated using a *and the* F *ratios:*

$$\lambda = \frac{a}{2} \cdot (F_1 + F_2 - 2) \qquad (10\text{-}3)$$

If our example had been an independent groups design, then $\lambda = (12/2) \cdot (72.07 + 32.30 - 2) = 614.22$. λ *is used to compute* a′ *in accord with the formula*

$$a' = \frac{(a + \lambda)^2}{(a + 2\lambda)} \qquad (10\text{-}4)$$

In this case, then, a′ *would be* $= 316.14$. *For an independent groups design,* a′, *rather than* a, *is used along with* b *to enter the table and obtain* W_0. *SCHUBRAD handles these variations automatically. If the F ratios in the example had come from an independent groups design, the critical value of* W_0 *would have been 1.32.*

Evaluation of the Schumann and Bradley Test

The Schumann and Bradley test has scarcely been used in research at all.[2] My guess is that this neglect is the result of ignorance; researchers do not use the test because it does not appear in textbooks. Yet within the limited context of its applicability, the test has considerable appeal. It is completely objective and can be carried out in a mechanical fashion; it is like an ANOVA on a pair of ANOVAs. If it is important to determine which of two similar experiments is more sensitive, the Schumann and Bradley test is ideal. The word "experiment" need not be taken literally. It may mean experimental method, as in the example; it may mean experimental subject, if the researcher wants to compare persons. It may even mean laboratory, if the researcher wants to contrast researchers.

Clinical Significance

The everyday usage of the term *significance* does not correspond well to the technical measures of magnitude of effect discussed above. Rather, what laypersons usually mean when they speak of the significance of a result corresponds most closely to what physicians refer to as clinical significance. Consider, for example, the case of a new drug hailed for its superiority in extending the lifespan of patients with a chronic, ultimately fatal, disease. The new drug might be significantly superior to its predecessor in the statistical sense, but if the extra longevity were typically only on the order of a week, standard medical practice would probably not be altered. The amount of improvement would be too small to be considered noteworthy, even if the difference between drugs is reliable.

In the behavioral world, the clinical significance perspective may be used to evaluate a statistically significant difference whose import is problematic. Weiss and Black (1995) found a significant difference between the extents to which male and female subjects endorsed rape myths. However, the mean difference was only half a point on a seven-point Likert scale. The argument was made that a difference corresponding to 7% of the scale represented basic agreement between men and women in a context in which a large discrepancy might well have been expected. Note that the argument focuses on the magnitude of the difference in relation to the scale, not in relation to variability.

Clearly, clinical significance is subjective. What is needed is an objective procedure that can take into account the utility of an observed difference. While statistical significance seems necessary before a difference can be declared clinically significant, substantive, domain-specific knowledge must be added to complete the assessment.[3] Suppose a study found that psychology graduates earned significantly more money five years after receiving their degrees than graduates in other sciences. Sounds good, doesn't it? However, if the mean difference were only $2 per week, then financial prospects would probably not be an important element in one's career decision.

It is unfortunate that the word *significant* was usurped by statisticians. Statistical significance is an important idea, one that deserves its own appellation.[4] The shared terminology has confused us all.

Post-hoc Power

When a test low in power reports no significant difference, the conclusion is virtually meaningless. Post-hoc power has been proposed by Keppel (1991) as a way to quantify the power of a completed experiment. Power is expressed as a probability, so the closer the reported number is to 1, the more powerful was the test. Generally, power below .8 would lead a researcher to conclude that a nonsignificant effect should be seen as inconclusive evidence.

Post-hoc power depends upon the degrees of freedom for numerator and denominator, the significance level, and the effect size. Computation has traditionally been carried out using the Pearson-Hartley charts (Pearson & Hartley, 1951) or with Cohen's (1977) tables, but these methods are cumbersome. It is preferable to invoke the post-hoc power option offered with the Anova table in the ONEWAY and FACTORIAL ANOVA programs; in FACTORIAL ANOVA, click the Hays, P-h P option available on the ω^2 button. For small experimental designs, such as those characterizing the problems in this text, post-hoc power is inevitably low.

The post-hoc power option comes with a consumer protection warning. Statisticians have questioned its utility (Hoenig & Heisey, 2001; Lenth, 2001), arguing that the calculated value is unlikely to tell the researcher anything that isn't already known. Customary usage is for a researcher who has failed to find an anticipated significant difference to attribute the disappointing result to insufficient power. Because the null hypothesis is never strictly true, confirmation via examination of post-hoc power merely reiterates what the significance test has already reported, namely that there was too little power to find the effect.

Exercises

10-1. To test the adage that fish is brain food, I arbitrarily divided my introductory psychology class into two groups. Half of the students were pledged to eat no meat during the quarter, and also to eat fish at least once per day. The other half were to eat no fish during the quarter. The scores on the final exam were the measure of performance. Determine whether the fish-eaters did better in the course, then estimate the degree of statistical association for the diet variable. Also examine the post-hoc power of this experiment, and decide how to evaluate the results.

Fish	No fish
86, 76, 62, 84, 78, 92, 95, 58, 47, 71	92, 83, 43, 61, 52, 58, 74, 68, 65, 57

10-2. Students are admitted to the training program at the Rorschach Training Institute only if they show sensitivity to differences in individual responses to the inkblots. The theory behind this policy is that the instructors can teach the syndrome labels that are associated with particular protocols, but the ability really to listen to what people say cannot be taught. Each prospective student is asked to rate protocols for degree of psychopathology on a 0–100 scale. Eight different protocols are each administered three times in random order. An applicant is admitted only if these ratings do not show significantly less sensitivity to the differences in the protocols than a similar set of ratings made by Dr. Einbrain, the institute's founder. Determine whether eager John Bluto will be admitted.

Ratings	Protocols							
	1	2	3	4	5	6	7	8
Dr. Einbrain	78	47	15	25	64	55	5	33
	63	53	17	28	59	47	8	42
	84	49	22	26	68	53	5	25
John Bluto	53	35	28	61	91	21	32	71
	71	42	22	58	80	15	45	65
	48	39	34	47	87	24	31	77

10-3. *Modern Psychology Magazine* recently published a Life Satisfaction Index; the higher a person's score, the more satisfying the life. I administered the test to 24 young psychologists, 12 men and 12 women. Half of each group had studied *Analysis of Variance and Functional Measurement*, and the other half had read other leading texts. Given the satisfaction scores below, estimate the proportion of variance associated with each of the variables. Also examine the post-hoc power for the three substantive effects.

	ANOVA and FM	Other texts
Men	73, 85, 57, 64, 92, 69	35, 61, 54, 70, 47, 65
Women	82, 59, 46, 53, 71, 80	71, 28, 52, 54, 48, 41

10-4. A developmental psychologist suggested that if a child is made responsible for the welfare of another living object, a sense of competence will develop that carries over into other arenas. This idea generated a research project in which fourth-graders who had experienced difficulty in third-grade math were the participants. At the beginning of the term each of the twenty children in the "slow-math" group was randomly assigned to one of four groups. Those in Group 1 were given a potted plant, and those in group 2 were given a kitten. Children in Group 3 received both a plant and a pet, while Group 4, a control group, got neither. The effectiveness of the manipulation was measured via midterm math scores. Use the partial ω^2 statistic to assess independently the impact of the plant, the kitten, and the interaction.

Group 1 (plant)	Group 2 (kitten)	Group 3 (both)	Group 4 (neither)
61	82	71	74
72	63	59	48
52	55	88	52
57	73	75	65
65	62	81	53

Answers to Exercises

10-1.

Source	df	SS	MS	F
Diet	1	460.80	460.80	2.03
Within groups	18	4083.00	226.83	

$\omega^2 = .049$

Post-hoc power is .16, so one cannot conclude diet doesn't matter.

10-2.

	Source	df	SS	MS	F
Einbrain	Protocols	7	11652.63	1664.66	52.29*
	Within cells	16	509.33	31.83	
Bluto	Protocols	7	10560.96	1508.71	30.33*
	Within cells	16	796.00	49.75	

$$W = 1.72$$

$$a' = 73.18; b = 8$$

$$W_0 \text{ (one-tailed, .05)} = 2.47$$

10-3.

Source	df	SS	MS	F	Post-hoc power
Gender	1	315.38	315.38	1.66	.08
Texts	1	1751.04	1751.04	9.22*	.49
GT	1	5.04	5.04	<1	0
Within cells	20	3797.50	189.88		

$$\omega^2_G = .021$$

$$\omega^2_T = .258$$

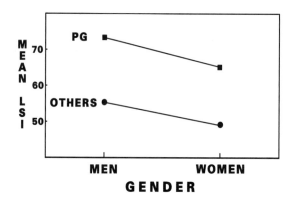

10-4.	Source	df	SS	MS	F	partial ω^2
	Plant	1	145.80	145.80	1.44	.02
	Kitten	1	605.00	605.00	5.97*	.20
	PK	1	28.80	28.80	<1	−.04
	Within groups	16	1621.20	101.33		

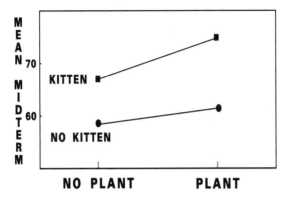

Notes

1. Techniques for obtaining ω^2 in more complex designs have been given by Vaughan and Corballis (1969).

2. The Schumann and Bradley test is used routinely in CWS applications, where the question addressed is whether one candidate judges more expertly than another. As discussed at the end of chapter 5, CWS captures expertise using the ratio of discrimination to inconsistency, often with mean squares providing estimates of these quantities. For other applications of the test, see Weiss (1980a) and Weiss (1986).

3. Clinicians have attempted to provide guidelines for assessment; a summary is given by Jacobson, Roberts, Berns, and McGlinchey (1999). My own somewhat extreme view, that subjective evaluation by an expert is not such a bad thing, is presented in Weiss, Edwards, and Weiss (2005).

4. Anderson (2001) has proposed the term *statsig*, which captures the idea nicely. I fear that the term is too inelegant to catch on.

11

Nested Designs

The term NESTING refers to a hierarchical arrangement among factors. Until now, we have considered only designs in which the factors are connected in the symmetrical relationship called crossing; that is, each level of each factor is paired with each level of every other factor. In a design with nesting, on the other hand, each level of the nested factor is paired with only one level of the superior factor. A superior factor is not in any sense better, of course; it is merely one under which another factor is nested. This terminology is perhaps less confusing than "nestor" and "nestee." Nesting is used when it does not make experimental sense to cross factors completely.

One kind of nested design may already be familiar. The design referred to as "mixed" or "within-between," is an example of nesting in which subjects are nested under one or more treatments. The mixed design has the character of repeated measures in that each subject is measured more than once but differs from the repeated measures designs we have seen previously in that the subject does not serve in all of the conditions. One reason the mixed design is used is that it may be impossible to have a person appear in all of the levels of one of the experimental factors. A factor that classifies people, such as gender, has this character. Few subjects are willing to undergo surgery for the sake of a fully crossed design. Because demographic variables are frequently incorporated into experimental research, the mixed design is a popular one. However, I don't like the imposition of a separate label. The mixed design is simply one of the possible nested designs, one in which subjects is the nested factor. An even worse label has entered the field via agronomy; the term *split-plot design* is also used for cases in which subjects (plots of land in the agricultural context) are nested under a combination of treatments. The procedures we develop apply to all cases in which one or more factors are nested; no special rules apply to the mixed design.

Let's delve into the nesting relationship with an example. Suppose we want to evaluate the usefulness of a newly discovered vitamin in helping neurologically handicapped children to remember more effectively. Memory performance may be evaluated in terms of performance on a standard digit span test. However, we cannot simply give the children the vitamin and test them because the effect of the vitamin may be evident only on a long-term basis. Therefore, we have to consider the environment of the children as well. Assume the children to be studied are in various institutions. A possible way to conduct the study would be to randomly assign the vitamin to some children and a placebo to others, but that might not be feasible in practice. For example, the administrator of an institution might not allow differential treatment of his charges. In formal terms, crossing vitamins and institutions would not be permitted.

A practical alternative is to give the vitamin to all of the children in some institutions, with children in other institutions receiving the placebo. In this case, institutions are nested under treatments. The solution is not entirely desirable because uncontrolled effects of particular institutions may cloud the effect of the experimental treatment. Using several institutions is an attempt to circumvent this difficulty by generalizing over institutions; the tactic is sensible, but what if the vitamin has different effects in different environments? This (perhaps unlikely) result is an interaction, but it cannot be evaluated with the given design. It cannot be evaluated because testing an interaction requires scores for all the pairings between the two factors.

The vitamin project illustrates both the advantage and the disadvantage of the nested design. The advantage is that the design allows the inclusion of a factor in the study even though crossing is impossible; certainly the impact of the study is greater when the vitamin is seen to promote improvement in several institutions rather than just one. The disadvantage of the nested design is that interaction between the superior and the nested factors is not estimable. The "interaction" does not exist, and should not be graphed.

Structurally the nested design can be viewed as a dissected factorial. Crossed factors may be schematized as a rectangle:

$$\begin{array}{c|c|c|c|} & A_1 & A_2 & A_3 \\ \hline B_1 & & & \\ \hline B_2 & & & \\ \hline \end{array}$$

Nested factors can be represented as separate columns:

$$\begin{array}{cc} & A_1 \\ B_1 & \\ B_2 & \\ \end{array} \quad \begin{array}{cc} & A_2 \\ B_3 & \\ B_4 & \\ \end{array} \quad \begin{array}{cc} & A_3 \\ B_5 & \\ B_6 & \\ \end{array}$$

Notice that the levels of the nested factor, B, have subscripts that differ from column to column. A nested design resembles an ordinary factorial design; the difference is that new levels of the nested factor appear with each level of the superior factor.

To identify nesting, look at the factors on a pairwise basis. If each level of one factor appears in combination with each level of the other factor, then the two factors are crossed. But if each level of one factor appears in combination with only one level of the other, then a nesting relationship exists between the two factors. The direction of the relationship, specifying which is the nested factor, is determined from the combining pattern. Several (at least two) levels of the nested factor appear with each level of the factor under which it is nested.

Nesting relations can get quite complicated. There can be multiple nesting, in which one factor is nested simultaneously under two or more others; there is ordered, or hierarchical, nesting, in which A is nested under B, which is in turn nested under C. The FACTORIAL ANOVA program handles nested designs of all levels of complexity automatically. The user need only specify the pairwise nesting relationship(s), as shown below, and the program constructs the appropriate F ratios. Using this feature would be the recommended usual approach for processing a nested design.

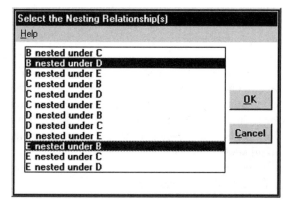

However, an understanding of these designs can in my view most effectively be gained by working through the structure and assembling the table yourself. The most convenient way to do this is by choosing the No F ratios option in FACTORIAL ANOVA's Design Specification window. Begin by pretending the factors are completely crossed. Construct the ANOVA table, omitting MSs and F ratios, and compute the SSs as usual. From this preliminary table, a corrected ANOVA table is put together. The corrections are based on POOLING the sums of squares and degrees of freedom of sources involved in nesting.

Fortunately, there is just one basic rule for determining how to pool sums of squares in nested designs, and this rule is applicable no matter how complex the design. After the rule is stated, we will work through its application to representative nested designs; these examples will also clarify the nature of nested designs.

The *fundamental rule for pooling* is that each nested source is pooled with its interactions(s) with the source(s) under which it is nested. Pooling denotes a summation of sums of squares and of the associated df. If a source is nested under two superior sources, then it is automatically nested under their interaction as well. It is convenient to write a replacement statement (computer programmers will recognize this structure), that is, a pseudo-equation, for each nested source. The replacement statement shows the pool that is used instead of the nested source. For example, if A *is nested under* B *and* C *is nested under* D, E, *and the* DE *interaction, then* two replacement statements are needed:

$$A = A + AB$$

$$C = C + CD + CE + CDE$$

Use of these replacement statements is straightforward. Every time a nested source appears in the preliminary table, whether by itself or as a component of an interaction, it is replaced by the corresponding right side of the pseudo-equation. For interaction terms, algebraic multiplication is used to determine the replacement pool. Using the example statements given above, the BC interaction would generate the pool $B \cdot (C + CD + CE + CDE) = BC + BCD + BCE + BCDE$. *Factors that become squared in the course of multiplication are canceled (i.e., set equal to 1 so that* $AB^2C = AC$).

This procedure works for all designs with nested factors, including those with repeated measures. The customary rules for determining error terms apply. In particular, it must be kept in mind that the error term for each source in a repeated-measures design is its interaction with subjects.

First let's look at the simplest kind of nesting. A school principal wants to compare two commercial reading programs. Two second-grade teachers are assigned to program 1 and two to program 2. The score is the amount of improvement shown by the child on the standard state reading test. For the example, we'll consider three scores from each class.

Program 1		*Program* 2	
Teacher 1	*Teacher* 2	*Teacher* 3	*Teacher* 4
16	14	9	7
10	15	10	9
12	11	8	7

The preliminary analysis calls for pretending the design is fully crossed. That means we regard teacher 3 as teacher 1, and teacher 4 as teacher 2; the design is a 2×2 with three scores per cell. The preliminary table:

Source	df	SS
Programs	1	65.33
Teachers	1	0.33
PT	1	3.00
Error (within groups)	8	32.00

To produce the corrected table, we analyze the nesting relationship. Here Teachers are nested under Programs. Therefore Teachers and PT are pooled to derive the corrected source, Teachers, which some researchers might label T/P or T(P) to symbolize the nesting. Both sums of squares and df are pooled as follows:

Source	df	SS	MS	F
Programs	1	65.33	65.33	16.33*
Teachers (T + PT)	2	3.33	1.67	<1
Error (within groups)	8	32.00	4.00	

Next, we consider a design in which subjects are nested under one of the two treatment factors. Three male fans and three female fans rated four sports announcers. The executive who commissioned the study was interested in comparisons between the announcers as well as whether males and females rated differently. The responses were converted to numbers on a 5-point scale.

Raters	Announcer			
	Howard	Curt	Tony	Joe
Male 1	1	3	4	5
Male 2	1	1	3	4
Male 3	1	2	4	5
Female 1	2	2	5	3
Female 2	1	2	5	2
Female 3	1	3	5	4

The preliminary table is computed as though the design is a $4 \times 2 \times 3$ complete factorial, with the raters appearing in both genders:

Source	df	SS
Announcers	3	38.79
Gender	1	0.04
Raters	2	3.00
AG	3	7.13
AR	6	1.33
GR	2	0.33
AGR	6	2.00

Because Raters is nested under Gender, we immediately pool Raters and the interaction between Raters and Gender (R = R + RG). Since there is more than one response from each subject, this design has the character of repeated-measures, with Raters being the name used for the subjects factor. As is usual in repeated-measures designs, the error term for each substantive source is its interaction with subjects. However, Raters will always be replaced by the pool as dictated by the replacement statement. Therefore, the error term for Gender can be seen to be the pool by following the usual composition rule to generate a Raters by Gender term,
RG = (R + RG) · G = RG + RG^2 = RG + R.

Announcers is not a nested source, so it is transferred to the corrected table intact. The error term for Announcers should be the interaction between Raters and Announcers, RA. Using the replacement again, RA = (R + GR) · A = RA + GRA. This pool is also seen to be the error term for the Announcers by Gender interaction, derived from the multiplication RAG = (R + GR) · AG = RAG + RAG^2 = RAG + RA.

Now we're ready to write the corrected table. It is convenient to write the error terms under the substantive sources that employ them.

Source	df	SS	MS	F
Gender	1	0.04	0.04	<1
Error (R + GR)	4	3.33	0.83	
Announcers	3	38.79	12.93	46.55*
AG	3	7.13	2.38	8.55*
Error (RA + RAG)	12	3.33	0.28	

The above is an example of what is (sometimes, and as I have explained, not by me) referred to as a mixed design. As can be seen, the rules for analysis are the standard ones.

Next we consider a more complex design, one in which there is double nesting. Volunteers who had been previously classified as either high- or low-anxiety persons were administered a drug. Half of the subjects were given saline, which is thought to have no effect, and the others were given a hormone that should augment the functioning of the autonomic nervous system. The researcher measured the pulse rate of the volunteers as they went through four stress conditions in an independently randomized order for each subject. Primary concern was with the interaction between anxiety and drugs; the main effects of these two variables were also of interest.

148 ANALYSIS OF VARIANCE AND FUNCTIONAL MEASUREMENT

			Stress Condition			
Anxiety	*Drug*	*Volunteer*	*Relaxation*	*Yelling*	*Isolation*	*Shock*
High	Hormone	1	61	82	79	95
		2	82	91	87	102
		3	73	78	79	89
	Saline	4	59	65	63	75
		5	64	72	65	78
		6	57	59	64	69
Low	Hormone	7	59	63	57	68
		8	62	65	64	71
		9	51	58	53	64
	Saline	10	52	61	54	63
		11	54	54	51	60
		12	53	57	50	62

In this example, anxiety, drug, and stress conditions are all fully crossed. Subjects are nested under drug and also under anxiety. The easiest way to recognize the nesting is to notice that although the design is laid out as though it were a complete four-way factorial, each volunteer appears in only one combination of anxiety and drugs. Incidentally, can you see why the label "volunteers" is used in the table? The reason is to avoid confusion in the ANOVA table between stress ("S") and subjects ("S"). Flexibility in labeling is worth developing to avoid ambiguous abbreviations.

The preliminary, as-if table is prepared first:

Source	df	SS
Volunteers	2	358.17
Anxiety	1	3040.08
Drug	1	1541.33
Stress	3	1309.75
VA	2	110.16
VD	2	117.16
VS	6	53.50
AD	1	432.00
AS	3	144.08
DS	3	47.83
ADS	3	24.83
VAD	2	30.50
VAS	6	76.17
VDS	6	8.17
VADS	6	56.16

Then the pooling rules must be invoked to determine the correct analysis. The nested factor is Volunteers, so every "V" in the table will enter into a pool. The pool involving Volunteers itself will also contain the interactions between Volunteers and its superior sources (that is, VA and VD), as well as the interaction between Volunteers and the joint interaction of the superior sources (VAD).

Each of the nonnested sources is transferred to the corrected table, and each of them requires an error term. It will be seen that each error pool serves several sources, but it is conceptually convenient to derive the error term for each source separately.

For A, we want VA. Using the replacement for V, we obtain $(V + VA + VD + VAD) \cdot A = VA + VA^2 + VDA + VA^2D = VA + V + VDA + VD$. *For D, we get* $VD = (V + VA + VD + VAD) \cdot D = VD + VAD + VD^2 + VAD^2 = VD + VAD + V + VA$. *For AD, the rule leads to* $VAD = (V + VA + VD + VAD) \cdot AD = VAD + VA^2D + VAD^2 + VA^2D^2 = VAD + VD + VA + V$. *Notice that in all three of the instances the elements in the pool are the same; therefore, the three sources utilize a common error term.*

Similar algebra is used to obtain the error terms for the other sources. For S, we get $VS = (V + VA + VD + VAD) \cdot S = VS + VSA + VSD + VSAD$. *The rule may be followed mechanically to yield the corrected table below.*

The corrected table is much shorter than the preliminary one. Again it is convenient to associate sources employing a common error term.

Source	df	SS	MS	F
A	1	3040.08	3040.08	39.48*
D	1	1541.33	1541.33	20.02*
AD	1	432.00	432.00	5.61*
Error (V + VA + VD + VAD)	8	616.00	77.00	
S	3	1309.75	436.58	54.01*
AS	3	144.08	48.03	5.94*
DS	3	47.83	15.94	1.97
ADS	3	24.83	8.28	1.02
Error (VS + VAS + VDS + VADS)	24	194.00	8.08	

For our last example, we consider a case of ordered nesting. Researchers in California and New York jointly performed a study measuring the cooperativeness of shopping center patrons using a "lost letter" technique. On each of the two days the researchers scattered 300 addressed letters through the parking lot; the letters were addressed to postal boxes and the number mailed in was the dependent variable. The researchers compared a small city to a large city in each state and contrasted a shopping center in a ghetto in each city to one in a nearby suburb. Also of interest to the researchers was a comparison of the addresses in each city. One hundred of each set of letters were addressed to the United Way, another hundred to the Chamber of Commerce, and the third hundred to the Department of Welfare. The scores in the table are the number of letters returned.

			Addressee		
State	City	Location	A_1(United Way)	A_2(CC)	A_3(Welfare)
S_1 (California)	C_1 (Los Angeles)	L_1 (Suburb) L_2 (Ghetto)	12, 31 20, 18	15, 23 3, 5	10, 12 21, 15
	C_2 (San Francisco)	L_3 (Suburb) L_4 (Ghetto)	39, 23 23, 51	17, 33 8, 2	24, 18 17, 21
S_2 (New York)	C_3 (New York City)	L_5 (Suburb) L_6 (Ghetto)	19, 25 27, 8	25, 12 5, 4	3, 2 27, 19
	C_4 (Buffalo)	L_7 (Suburb) L_8 (Ghetto)	41, 29 29, 33	31, 19 11, 4	6, 17 25, 31

This design looks like a 2 × 2 × 2 × 3 factorial with two scores per cell. The preliminary analysis proceeds on this superficial basis:

NESTED DESIGNS 151

Source	df	SS
State	1	1.69
City	1	760.02
Location	1	72.52
Addressee	2	1515.04
SC	1	1.69
SL	1	46.02
SA	2	6.13
CL	1	13.02
CA	2	193.29
LA	2	1480.29
SCL	1	0.19
SCA	2	2.38
SLA	2	249.04
CLA	2	67.54
SCLA	2	22.63
Error (within cells)	24	1499.50

Then the nesting must be untangled. Here it is ordered; Location is nested under City (each particular shopping center is in only one city, and such centers are unique enough that we would regard those in different locales as distinct), while City is in turn nested under State. This hierarchical structure implies that Location is also nested under State. The corrected table is constructed in accord with the nesting rules. Since City is nested under State, the corrected source for City is the result of pooling C + SC.

For Location, we have the pool L = L + CL + SL + SCL, because Location is nested under City and State and therefore under their interaction. Only three sources are not nested and are therefore transferred intact; they are State, Addressee, and their interaction. Two distinct interaction pools can also be estimated. The AC interaction is A·(C + SC), while the AL interaction is A·(L + CL + SL + SCL) = LA + CLA + SLA + SCLA.

Finally we can write the corrected table. Summing the degrees of freedom provides a check that no terms have been lost. Also it is worth verifying that every source in the original table appears exactly once in the final table, either by itself or in a pool.

Source	df	SS	MS	F
State	1	1.69	1.69	<1
City (C + SC)	2	761.71	380.85	6.10*
Location (L + CL + SL + SCL)	4	131.75	32.94	<1
Addressee	2	1515.04	757.52	12.12*
SA	2	6.13	3.06	<1
CA + SCA	4	195.67	48.92	<1
LA + CLA + SLA + SCLA	8	1819.50	227.43	3.64*
Error (within cells)	24	1499.50	62.48	

As we worked our way through the jungle of nested designs, a light may have dawned on you. The structure of the nested design was not wholly unfamiliar. In fact, it has been with us all along, albeit in disguised form. In an ordinary factorial design with several scores per cell, replicates is a nested factor. In fact, it is simultaneously nested under all of the other factors. Applying the nesting rule for pooling leads to a single pool that is ordinarily used as the error term. Let us verify this by tracking through an example. Assume that each score comes from a separate subject.

		B		
		1	2	3
A	1	72, 83, 63, 48, 52, 90	27, 42, 64, 39, 72, 48	89, 94, 63, 99, 87, 92
	2	92, 63, 45, 57, 59, 26	67, 55, 80, 22, 71, 66	91, 82, 78, 98, 100, 76

Here the factor we call replicates has six levels. The preliminary, as-if, analysis is of a three-factor design with no obvious error term:

Source	df	SS
R	5	742.89
A	1	0.44
B	2	7100.72
RA	5	1442.89
RB	10	4189.94
AB	2	759.39
RAB	10	2191.28

Replicates is nested under both A and B, and therefore is automatically nested under their interaction. So the corrected source for Replicates is the pool consisting of Replicates plus the interaction of Replicates with A, B, and AB. Thus the corrected table is

Source	df	SS	MS	F
A	1	0.44	0.44	<1
B	2	7100.22	3550.36	12.43*
AB	2	759.39	379.69	1.33
R (R + AR + BR + ABR)	30	8567.00	285.57	

If you compare the sums of squares from this table with those of the first numerical example in chapter 5, the nature of within-cells error will be apparent. The logic exemplified here is not limited to two-factor designs; the within-cells error consists of every source in which an R appears, whatever the number of crossed factors.

Simple Effects in Nested Designs

When there are two or more substantive factors, it may be of interest to examine a pairwise interaction by looking at the simple effects. If the pair consists of a substantive nested factor and a factor under which it is nested, simple effects are not defined because the interaction cannot be estimated. If the subject factor is nested, computational agony arises because the error term for the simple effect is complex. Rather than present the alternative formulation, I recommend that you let FACTORIAL ANOVA carry out the computations.

A Note on Change Scores

In studies in which learning is a goal, the status of each individual at the end of the project may depend on how much was known at the beginning. Therapeutic intervention may have the same character in that initial health may be a determinant of final health. If patients are randomly assigned to treatments, the researcher has the option of ignoring initial status and using only the final score, as was recommended by Cronbach and Furby (1970). This course would seem especially sensible if assessment itself might have an impact, so that it would be preferable to measure only once. On the other hand, if there were reason to anticipate high correlation between pretest and posttest scores, one would expect a more powerful analysis with interindividual differences removed.

One way to accomplish the removal is to carry out a repeated-measures ANOVA with participants nested under the treatments and crossed with the two times. The F ratio for the treatment by time interaction evaluates the null hypothesis that change has been equal across treatment conditions. This F ratio is exactly equivalent to the one obtained from a one-way ANOVA on the change scores (subtracting pretest from posttest yields the score for each person), a mathematical fact noted by Huck and McLean (1975) and proved by Maxwell and Howard (1981). In general, though, analysis of covariance is a preferable technique for dealing with change scores because it affords greater power in most situations (Huck & McLean, 1975).

Exercises

11-1. Six randomly selected subjects were tested in an experiment on the effects of sunglasses on visual acuity. The tests were conducted using all possible combinations of two types of sunglasses and three different lighting conditions. Each subject was tested under one of three levels of initial light adaptation. The score given is a measure of acuity, with higher numbers representing better acuity. Evaluate these data. Find the simple effect of sunglass type for each initial adaptation.

154 ANALYSIS OF VARIANCE AND FUNCTIONAL MEASUREMENT

		Initial adaptation						
		1			2		3	
		Sunglass type			Sunglass type		Sunglass type	
Lighting	Subject	1	2	Subject	1	2	Subject	1	2
L_1	S_1	10	8	S_3	50	53	S_5	37	42
	S_2	6	11	S_4	47	61	S_6	43	50
L_2	S_1	14	9	S_3	33	42	S_5	25	28
	S_2	10	8	S_4	48	55	S_6	30	32
L_3	S_1	0	10	S_3	37	62	S_5	24	27
	S_2	11	16	S_4	49	60	S_6	21	34

11-2. A psychologist wanted to examine the effect of a special maze-training program she had developed for mice. The training consisted of putting the mice on carts and pulling them through ten different mazes. She thought this training would help mice learn hard mazes more quickly but would not have much effect on learning easy mazes. Because she had read that different strains of mice, which can become highly inbred by the supplier, are sometimes "maze-bright" or "maze-dull," she bought her animals from three mouse suppliers. The trained mice were put on the training carts at age 3 weeks, while the control mice were put on similar carts but were simply pulled back and forth. Each animal was then tested on the easy and hard mazes, and the number of trials until error-free performance was achieved was recorded. Half of the animals did the easy maze first. Analyze the data.

		Trained		Control	
		Easy maze	Hard maze	Easy maze	Hard maze
Supplier A	Easy first	8	14	15	22
		6	11	18	19
		5	12	14	21
	Hard first	9	16	16	24
		8	13	17	21
		10	14	19	28

NESTED DESIGNS 155

		Trained		Control	
		Easy maze	Hard maze	Easy maze	Hard maze
Supplier B	Easy first	12	13	18	26
		14	17	14	28
		8	14	13	19
	Hard first	15	23	21	23
		10	18	19	41
		13	15	16	28
Supplier C	Easy first	11	18	19	28
		9	22	23	22
		14	19	17	19
	Hard first	14	22	21	33
		12	14	26	40
		10	13	31	29

11-3. A study of chimpanzee language behavior was concerned with how many words the chimp would sign (using American Sign Language) under various field conditions. Because each chimp had been assigned to a specific trainer, it was also of interest to examine whether the trainers were equally effective. The score is the number of words spontaneously signed by the chimp during a half-hour hour test. Analyze the data.

		Field conditions		
Trainer	Chimp	Bare room	Kindergarten	Artificial jungle
Roger	Lucy	10	31	15
	Allee	14	40	11
	Bill	6	23	17
Duane	Sarah	32	49	38
	Lana	4	7	8
	Pooh	23	30	22
Bea	Washoe	40	75	63
	Jerry	35	49	41
	Jim	27	28	34

11-4. A professor of communications assigned students to count the number of killings shown on a specific channel during two specific hours on Thursday night. To reduce the likelihood of recording errors disrupting the research, two students were assigned to each period. All students watched alone. The professor was interested both in comparing networks and in whether the broadcasters respected the "family hour," 8 to 9 p.m. The scores are the number of killings reported. Analyze the data. Find the simple effects of time for each channel.

	Channel		
Time	2	4	7
8–9 p.m.	Joe 4 Bill 4	Sally 5 Leo 5	Pat 15 Ann 15
10–11 p.m.	Joe 11 Bill 12	Sally 7 Leo 7	Pat 17 Ann 15

11-5. Four clinical psychologists and four experimental psychologists each rated the quality (on a 6-point scale) of the research reported in journal articles. Three articles were randomly chosen from a clinical psychology journal, and three others were chosen randomly from an experimental journal. Analyze the ratings.

			Clinical journal			Experimental journal		
			Article A	Article B	Article C	Article D	Article E	Article F
Clinical psychologists		RC	4	3	4	2	3	2
		HG	6	5	4	4	3	3
		RL	5	5	4	3	5	3
		BS	4	3	3	3	2	2
Experimental psychologists		DP	1	2	2	5	5	4
		DF	2	1	3	4	4	5
		CG	3	2	2	6	5	5
		BL	1	1	1	2	4	3

Answers to Exercises

11-1.

Source	df	SS	MS	F
Adaptation	2	9840.89	4920.44	65.63*
Error (Sub + SA)	3	224.92	74.97	
Type	1	354.69	354.69	74.67*
AT	2	142.89	71.44	15.04*
Error (ST + STA)	3	14.25	4.75	
Lighting	2	328.72	164.36	6.39*
AL	4	482.94	120.74	4.69*
Error (SL + SLA)	6	154.33	25.72	
TL	2	121.06	60.53	3.21
TLA	4	18.61	4.65	<1
Error (SLT + SLTA)	6	113.00	18.83	

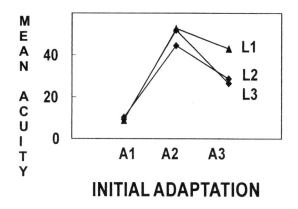

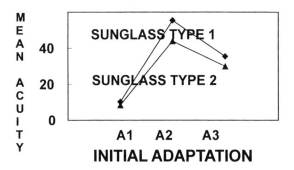

158   ANALYSIS OF VARIANCE AND FUNCTIONAL MEASUREMENT

Simple Effects	df	SS	MS	F
Type at A_1	1	10.08	10.08	4.84
ST at A_1 (error)	1	2.08	2.08	
Type at A_2	1	396.75	396.75	190.44*
ST at A_2 (error)	1	2.08	2.08	
Type at A_3	1	90.75	90.75	9.00
ST at A_3 (error)	1	10.08	10.08	

11-2.

Source	df	SS	MS	F
Training	1	1530.89	1530.89	108.73*
Supplier	2	337.00	168.50	11.96*
Order	1	200.00	200.00	14.20*
TS	2	27.44	13.72	<1
TO	1	80.22	80.22	5.70*
SO	2	6.33	3.17	<1
TSO	2	78.11	39.06	2.77
Error	24	338.00	14.08	
(Mice + MS + MO + MT + MTS + MTO + MSO + MTSO)				
Difficulty	1	760.50	760.50	63.38*
DT	1	16.06	16.06	1.34
DS	2	12.33	6.17	<1
DO	1	6.72	6.72	<1
DTS	2	38.78	19.39	1.62
DSO	2	4.78	2.39	<1
DOT	1	16.06	16.06	1.34
DTSO	2	14.78	7.39	<1
Error	24	288.00	12.00	
(MD + MDT + MDS + MDO + MDTS + MDSO + MDOT + MDSOT)				

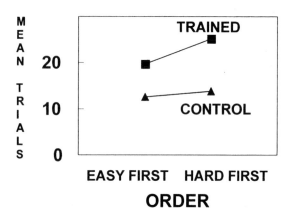

11-3.

Source	df	SS	MS	F
Trainers	2	3140.07	1570.04	3.07
Error (Chimp + CT)	6	3071.11	511.85	
Field	2	1116.07	558.04	12.86*
TF	4	216.37	54.09	1.25
Error (CF + CTF)	12	520.89	43.41	

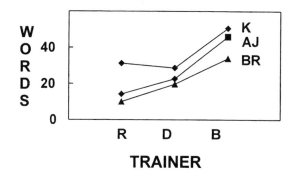

11-4.

Source	df	SS	MS	F
Channels	2	204.50	102.25	245.40*
Error (Students + SC)	3	1.25	0.42	
Times	1	36.75	36.75	88.20*
CT	2	24.50	12.25	29.40*
Error (ST + SCT)	3	1.25	0.42	

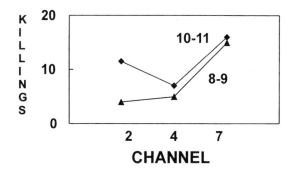

Simple Effects	df	SS	MS	F
Time at Channel 2	1	56.25	56.25	225.00*
ST at Channel 2	1	.25	.25	
Time at Channel 4	1	4.00	4.00	undefined
ST at Channel 4	1	0	0	
Time at Channel 7	1	1.00	1.00	1.00
ST at Channel 7	1	1.00	1.00	

Note: The program warns of possible errors when the denominator of an F ratio is zero, because that usually suggests faulty input. Here, though, there really is no variability in some of the error terms (look at the data).

11-5.

Source	df	SS	MS	F
Type of Psychologist	1	3.00	3.00	<1
Error (Subjects + ST)	6	19.92	3.32	
Journals	1	5.33	5.33	20.21*
JT	1	44.08	44.08	167.05*
Error (SJ + STJ)	6	1.58	0.26	
Articles (A + AJ)	4	2.08	0.52	1.04
TA (TA + TAJ)	4	1.92	0.48	<1
Error (SA + STA + SAJ + STAJ)	24	12.00	0.50	

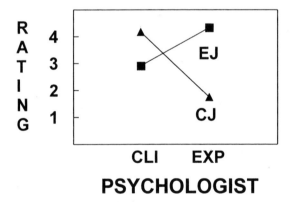

12

Missing Data

Every now and then the best laid plans of rats and sophomores go awry, and a carefully designed study ends up with missing scores. An animal may die, or a human may not show up, and the researcher must deal with a reduced set of data. In terms of the statistical analysis, this reduction can be either that the various cells in the design no longer have the planned equal numbers (the unequal n case) or that one or more cells in the design have no scores at all (the missing cells case). Also, we shall consider the rare situation in which inequality of cell size is intentionally achieved.

Independent-Groups Designs

The situation we examine first is one in which each cell has some scores in it, but the number of scores differs from cell to cell. While such unequal n is of little consequence for one-way designs, special consideration is required for factorial designs. The researcher should not blindly submit the data to a computer program, even one that claims to cater to the situation.

Any method for coping with unequal cell sizes requires a critical assumption about the nature of the missing scores. The most comfortable assumption is that the missing scores have occurred randomly. More formally stated, this assumption is that there is no connection between the treatments and the particular data values that are missing. For example, if fewer volunteers show up for the high-voltage condition, or if more animals die after receiving a particular drug or surgical treatment, the loss of scores could hardly be viewed as random. In such cases, the obtained number of scores reflects true aspects of the treatment, and the approach to the data must incorporate that fact.

A Test of the Randomness Assumption

The decision that missing scores have occurred systematically should not be made lightly, for its consequence is likely to be rejection of the validity of the data set. While the reasons underlying particular missing scores cannot generally be known, widely disparate patterns in the cell sizes suggest that biased selection may have occurred. A formal test of the hypothesis that the number of scores differs significantly across conditions can justify the drastic step of refusing to accept the data at face value. There will also be occasions when attrition is an interesting outcome in its own right.

An ANOVA that considers the presence or absence of a score, rather than its magnitude, fills the bill (Weiss, 1999). The procedure is simple and uses familiar methods. The null hypothesis is that the number of lost scores for all design cells is equal. Attending to the factorial structure, the analysis calls for replacing each actual score with a 1 and each planned but missing score with a 0. This ANOVA will therefore have equal cell n, and that n (the planned cell size) will be the number of replicates. The values 1 and 0 are arbitrary, as any other values would yield the same F ratio; but 1 and 0 form a natural code for "present" and "absent." The Random Attrition option available via the Unequal Cell Size menu presented in the opening window of FACTORIAL ANOVA may be used to carry out this test. A significant F ratio for any substantive source marks concentrated inequality of attrition. Specific comparisons may also be used when the researcher anticipates that particular treatment combinations may prove troublesome. The present-absent coding could also be used to test for systematic trend in the pattern of missing scores. Here is the setup to generate the Attrition ANOVA table (for exercise 12-4); this grid becomes available after the parameters have been entered in the opening window. The data entered are the number of participants in each condition.

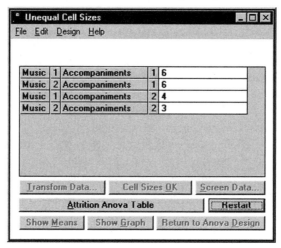

Researchers also face unequal cell sizes when data have not been lost in a literal sense. Independent groups designs are used in cases in which participants have not been randomly assigned to groups, but instead have been recruited to fill particular design cells. Often the factors are personal characteristics not subject to randomization, such as gender or ethnicity. Cell inequalities reflect differences in ease of recruitment, an issue that may have substantive importance. Here the null hypothesis is that the number of scores present for all design cells is equal. The slight difference in wording leads to a difference in the way the test is carried out. It seems appropriate to consider as the number of replicates the largest number of participants obtained for any cell. The adjustment ensures that the researcher will not be "rewarded" for an overly optimistic projection of cell sizes, which might have the consequence that all cells would fall far short of the goal and the analysis would have little power to detect disparities.

Coping with Accidental Inequality

The simplest compensatory procedure is random deletion of the scores at hand to achieve equality of cell size. Suppose the design is a 2 × 2, and there are 33, 30, 36, and 34 scores in the cells. Whereas the original plan may have called for 36 scores per cell, haphazard circumstances beyond the researcher's control have produced a situation in which 30 is the maximum cell size if equal n is in sight. The researcher would randomly delete 3, 6, and 4 scores from the first, third, and fourth groups respectively and then proceed with the usual analysis. It is important to stress that deletion must be done randomly, without looking at the score values or deciding that certain subjects are apt to produce unrepresentative scores.

Throwing out data is not a desirable solution to the problem of unequal cell size. The procedure (*technique* seems too flattering a term) is sensible only when the smallest cell n is fairly large (20 is a practical minimum), and even then it seems wasteful. After all, if the data are inexpensive enough (in terms of either labor or cash) that one can afford to throw some away, then perhaps it would not be impractical to collect more data and bring the deficient cells up to equality. If a moralistic tone seeps through in this presentation, I must confess that painful memories of data collection have made me miserly about wasting responses.

Unweighted Means

The technique known as UNWEIGHTED-MEANS analysis is a more desirable solution to the problem of unequal cell size.[1] It is applicable to small group sizes as long as there is not great disparity among the ns, and it does not call for any scores to be omitted from the analysis. The name of the method stems from the property that each cell mean contributes equally (thus, it is not weighted for cell size) to the sum of squares. The basic assumption is that the obtained cell means reflect the means that would have occurred if no scores had been lost. Since this assumption is the natural one if we hold that missing scores have occurred randomly across conditions, it follows that the unweighted-means analysis is generally the method of choice for unequal cell sizes.

The unweighted-means analysis requires somewhat more effort than the standard ANOVA. It will be instructive to carry out the computations using familiar computer routines. The ONEWAY ANOVA and FACTORIAL ANOVA programs may be used; only a slight amount of additional manual labor is required. After learning the technique, we can use the Unweighted Means option, available through the Unequal Cell Size menu in the opening window of FACTORIAL ANOVA, for the complete analysis.

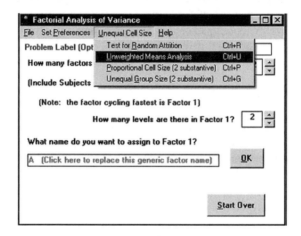

Here are some data from a 3 × 2 independent-groups design with unequal cell sizes:

		A		
		1	2	3
B	1	7, 3, 6, 6, 4, 2	8, 24, 21, 12, 17, 22, 16, 19	16, 8, 14, 15, 17
	2	23, 18, 14, 22, 9, 26	11, 31, 13, 15, 26, 26, 14	9, 42, 18, 27, 16, 20, 31, 17

It would seem obvious that there is no great disparity among the cell sizes, and the ANOVA test for cell size inequality (using 8 as the number of replicates) produces no significant F ratios.

Source	df	SS	MS	F
A	2	.292	.146	1.07
B	1	.083	.083	<1
AB	2	.542	.271	1.98
Within cells	42	5.750	.137	

To analyze the scores without a specialized program, we proceed in two stages. The first consists of analyzing the factorial design as though it were a one-way design. This means you must temporarily ignore whatever structure is in the design. The objectives of this step are to obtain the cell means and the within cells mean square that will be the error term (for these data, it is 47.18) for the analysis.

The second stage consists of a two-way ANOVA on the cell means. The analysis is performed in the ordinary manner except that there is no within-cells term computed. This omission is hardly surprising since there is only one "score" (really a mean) per cell. The means to be analyzed may be conveniently arranged in a table that has the same form as the table of raw data.

		\multicolumn{3}{c}{A}		
		1	2	3
B	1	4.67	17.38	14.00
	2	18.67	19.43	22.50

Source	df	SS (raw)	MS (raw)
A	2	59.15	29.58
B	1	100.48	100.48
AB	2	35.75	17.88

Before the final ANOVA table is written, the sums of squares and mean squares for the substantive sources (but not for the error term) must be adjusted. The adjustment consists of multiplying the terms by the harmonic mean ($\tilde{n}$) of the cell sizes. The harmonic mean is obtained by dividing the number of cells by the sum of the reciprocals of the individual cell sizes. In the present example, there are six cells (because it is a 3×2 design). The sum of the reciprocals of the cell sizes is:

$$1/6 + 1/8 + 1/5 + 1/6 + 1/7 + 1/8 = .926.$$

So $\tilde{n} = 6/.926 = 6.48$.

The final table is constructed, then, by multiplying the substantive MSs by $\tilde{n}$; we then bring in the error term and derive the F ratios. Omitting the sums of squares simply saves some irrelevant multiplication.

Because this analysis lacks orthogonality, the sum of the treatment sums of squares in the completed table will in general be unequal to the sum of squares for treatments obtained in the preliminary, one-way ANOVA. Here is the output obtained using the Unweighted Means option in FACTORIAL ANOVA:

Source	df	SS (raw)	MS	F
A	2	59.15	191.59	4.06*
B	1	100.48	650.92	13.80*
AB	2	35.75	115.81	2.46
Within cells	34	1604.26	47.18	

Proportional Cell Sizes—Intentional Inequality

In contrast to the unweighted-means approach, the analysis of the special case in which cell sizes are proportional incorporates weighting each cell by the number of cases in it. This weighted analysis is appropriate when inequalities in cell sizes reflect inequalities in the population. For example, if ethnic groups were a factor in the design, the researcher might choose to employ white, black, and Asian subjects in proportion to their representation in the population of interest. Another example might be one in which IQ or GPA was a factor, with the study including more subjects from intervals that are heavily represented on campus. The proportion in the experiment would be matched to the proportion in the population. An experiment may include one or more of these classificatory factors.

There are two options when the researcher wishes to make a classification a factor in the experimental design. Undoubtedly the more common choice is to have equal numbers of the various levels of the classification factor; for example, one might recruit an equal number of left-handed and right-handed volunteers. This option would be appropriate if there were substantive interest in the handedness element (i.e., does handedness have an impact, and does it interact with the other factors in the study?).

The alternative approach would make use of proportional cell sizes. This choice would be preferred if the researcher was not really interested in the classification factor *per se* but wanted to be sure that the subject groups accurately reflected population proportions. The worry would be that a typical handy sample of volunteers might not provide proper representation of the subgroups. Federal medical research guidelines, as well as those of the APA, urge researchers to include broad representation in study samples. The motivation for these policies is that, in the past, researchers have neglected to incorporate some segments of the population in experiments. This neglect weakens the generalizability of the findings. Use of the proportional cell size design allows the substantive variables of primary interest to be measured over a more representative sample. The impact of the classification variable would be assessed, but it would likely not be given full attention; if it proved significant, a full-scale follow-up with equal n might be in order.

The weighted-means analysis is used when inequality of cell sizes in the experiment is intentional. The researcher intends that estimates of the substantive effects depend upon proportions in the population. The population proportions are an interesting aspect of the situation. That is why a weighted analysis is chosen. On the other hand, if disparities in cell size were the results of random occurrences, estimates using a weighted analysis would depend upon proportions in the sample. Because sample proportions would be regarded as uninteresting in the case of accidental inequality, an unweighted analysis would be preferred.

Cell sizes are proportional when each cell size is the product of a proportionality constant for the row and another for the column. Either constant may be one. All of the sets of frequencies on each factor must be proportional to each other, and to the marginal frequencies for that factor. As an illustration, consider a two-factor design with cell n as indicated:

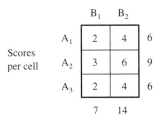

The proportionality can be seen for A: 2 is to 4 as 3 is to 6 and as 2 is to 4, and also as 7 is to 14. All of the pairs of numbers have the ratio 1 to 2. For B a similar situation obtains: 2 is to 3 is to 2 as 4 is to 6 is to 4, and also as 6 is to 9 is to 6. Here the triplets have the ratio 2 to 3 to 2. Notice that the cell ns can be determined by multiplying the ratios for the rows and columns. This is the defining characteristic of the proportional cell size design.

The complete analysis, for a design with two substantive factors, may be conducted using the Proportional Cell Size option available through the Unequal Cell Size menu in the opening screen of FACTORIAL ANOVA.

We will, as usual, illustrate the computations as they might be carried out without a specialized program. The example is a study conducted by a social psychologist exploring stereotypes. Each member of the psychology department faculty was given a picture of an applicant and asked to rate her chances of success in the graduate program on the customary 9-point scale. The pictures were color portraits from a college yearbook, chosen by the researcher to explore the variables of hair color and glasses in roughly natural proportions. Here are the ratings. Check to see that the cell ns are indeed proportional. (Notice that there are twice as many scores in corresponding cells for "no glasses.")

	Glasses	No glasses
Blonde hair	6, 4, 5, 7	4, 1, 3, 2, 6, 2, 3, 4
Brown hair	6, 9, 8, 3, 8, 7, 9	7, 3, 4, 5, 2, 3, 1, 6, 2, 4, 3, 5, 6, 3
Red hair	2, 6	3, 5, 1, 3

For the sake of illustration, we will proceed from scratch. As usual, we begin by computing $\Sigma X^2 = 937$ and $T^2/N = 171^2/39 = 749.77$. This arithmetic could be carried out with the ONEWAY ANOVA program. The calculations are based on cell totals, so we reduce the data accordingly:

	Glasses	No glasses	
Blonde hair	22 n = 4	25 n = 8	47
Brown hair	50 n = 7	54 n = 14	104
Red hair	8 n = 2	12 n = 4	20
	80	91	

The computations proceed as in the ordinary way, except that the totals that are squared and then summed do not share common divisors; rather each total is divided separately by the number of scores that went into it. Here are the manual calculations:

$$SS_{Hair} = 47^2/12 + 104^2/21 + 20^2/6 - T^2/N = 16.03$$

$$SS_{Glasses} = 80^2/13 + 91^2/26 - T^2/N = 61.04$$

$$SS_{H \times G} = 22^2/4 + 50^2/7 + 8^2/2 + 25^2/8 + 54^2/14 + 12^2/4$$

$$- T^2/N - SS_H - SS_G = 5.72$$

The F table can be filled in without any new wrinkles. The df for error may be determined via the rule that each score beyond the first in a cell provides one df for the within cells term.

Source	df	SS	MS	F
Hair	2	16.03	8.01	2.53
Glasses	1	61.04	61.04	19.29*
HG	2	5.72	2.86	<1
Within cells	33	104.45	3.17	

The weighted-means analysis is orthogonal, implying that you can check your arithmetic by comparing the sum of the treatment sums of squares (16.03 + 61.04 + 5.72 in the example) with the treatment sum of squares from a one-way ANOVA on these data; the two sums should be equal.

We have examined two methods for dealing with unequal cell sizes. The method of unweighted means is appropriate when scores are lost haphazardly, while the analysis of proportional cell sizes is appropriate for planned inequality that mirrors proportions in the population. Incidentally, algebra lovers can confirm that if the cell ns should happen to be equal, either of these approaches to unequal n data sets will reduce to the correct analysis. Both of these approaches can be applied to specific comparison problems as well.

Repeated-Measures Designs

Accidentally Missing Scores

When there is only one score per cell, a missing score implies an empty cell. There is no cell mean from which to estimate the missing score. The researcher must first decide if the empty cells have arisen through reasons unrelated to the treatments and can thus be regarded as accidental or random. If the number of missing scores is small and if accidents seem plausible explanations for them, then it may be appropriate to estimate the absent scores. A simple, sensible scheme has been given by Winer (1971). Each empty cell is estimated from its neighbors; thus the missing scores are estimated from those present.

Consider as an example a repeated-measures design in which each of four students was to take five exams. However, two students each missed one of the tests. The research interest is in determining whether the exams differ in difficulty. Here are the scores:

Students	1	2	3	4	5
A	53	62	47	73	81
B	x	58	39	63	75
C	41	57	35	59	x
D	62	74	68	52	91

The first problem is to estimate the missing scores. These estimates are then inserted into the data and an ordinary repeated-measures ANOVA is performed. One degree of freedom is removed from the interaction (that is, from the error term) for each score that was estimated.

The estimation method assumes the ratio of adjacent cells is the mean of the ratios of adjacent neighboring cells. Thus, an estimate of the missing value at B_1 may be given by the equation: $B_1/58 = 1/2[(53/62) + (41/57)]$. To solve the equation, isolate the unknown, B_1 (i.e., $B_1 = 58 \cdot 1/2 \{[53/62] + [41/57]\} = 45.65$). In a similar way, the missing value at C_5 is estimated: $C_5/59 = 1/2 [(75/63) + (91/52)]$. Solving for C_5 yields $C_5 = 86.74$. Inserting these two estimates leads to the ANOVA table:

Source	df	SS	MS	F
Students	3	633.21		
Exams	4	3232.16	808.04	9.78*
Error (SE)	10	826.51	82.65	

As the estimation produces a complete factorial design, the sums of squares (but not MS_{error}) may be furnished by FACTORIAL ANOVA.

It should be reiterated that the substitution of these estimated scores is appropriate only when a small number of isolated scores is missing and only when the missing scores have come about in a haphazard way. The presumption is that the missing scores are predictable from those present on the basis of local lack of interaction. If there are more than a few missing scores or if the missing scores are systematically associated with particular treatments, then the estimates are of dubious validity. It should go without saying that the estimation procedure given here is only a Band-Aid, and that real data are always preferable to estimates. It should also be apparent that the procedure given here allows a choice of which neighboring scores to use for the estimate, and the ANOVA will depend on which neighbors have been chosen. It is possible to eliminate this indeterminacy by using an estimation procedure based on all of the scores (a least-squares procedure). However, the additional arithmetic may not be justified, as it presumes the dubious assumption that distant scores are as relevant to the missing cell as are near neighbors.

More sophisticated estimation schemes are discussed extensively by Little and Rubin (1987). A summary of recently developed imputation schemes is given in Schafer and Graham (2002). Multiple imputation Schafer (1999) is a simulation-based approach in which each missing score is replaced by a number of simulated values. The simulation can incorporate the analyst's perspective on how the observed scores might contribute unequally to the estimates. The resulting versions of the data are then analyzed as though the data set were complete, and the outcomes are averaged to produce inferential statements that incorporate the uncertainty caused by the missing scores. These procedures are too complex to be carried out manually; specialized programs are available.[2]

Systematically Missing Scores

Unequal Group Sizes

When participants are recruited and run in naturally occurring groups, it is unlikely that the groups will be matched in size. For example, a researcher in education may wish to test a curricular innovation or compare textbook effectiveness. The practical way to apply the treatments is on a classroom basis, and of course the classes may well contain different numbers of students. If the program calls for more than one measurement per student, then participants must be a factor in the design. That factor is nested under the grouping variable, classrooms.

The nested source, participants, is crossed with the repeated-measures factor. Since each subject is in all of the treatments, cell sizes across the repeated-measures factor are equal. Therefore, the design has the character of proportional cell sizes, with the proportionality constants for the levels of the grouping variable given by the number of subjects in each group. The proportionality constants for the repeated-measures factor are all one. Proportional designs produce orthogonality and thereby generate additive sums of squares.

The setting for our example is Luddite Elementary School, an institution noted for small class sizes. A generous computer manufacturer donated several machines to the school. The principal decided to evaluate them by placing the machines in one of the third-grade classrooms, while the other third-grade classroom got none. The manufacturer claimed there would be visible impact in three distinct areas by semester's end. The teacher furnished the principal a confidential numerical assessment of each child's performance in three fields: math, English, and art.

Ms. Brooks's classroom
(computers)

	Math	English	Art
Sharon	80	92	75
Tracy	62	73	90
Kip	95	75	50
Richard	68	45	63
Jennifer	75	83	67
Sean	40	52	48
Tai	79	84	97
Barbi	50	45	58
Brook	95	68	82

Mr. Novak's classroom
(no computers)

Jeff	57	71	64
Tawana	47	85	92
Mari	62	58	74
Scott	49	61	55
Kimiko	73	59	72
Bob	44	52	68
Chantal	60	79	90

The FACTORIAL ANOVA routine for this design is accessed through the Unequal Cell Size menu in the opening window. The program accommodates designs in which subjects are nested under one factor and crossed with another.

To understand how the analysis works, we can go through the calculations without using the specialized routine. We partition the data several times and do ordinary ANOVAs, then invoke the usual nesting rules and carry out pooling manually. Understanding the method is not only valuable in its own right, but will allow us to construct the correct analysis for more complex designs. Sums of squares and dfs are the quantities that are additive, so it is those with which we shall work. To begin the computations, ignore the grouping factor, classrooms; enter the data as though the design were simply sixteen subjects by three treatments. From this analysis, we get SSs for "Students," with 15 df (= 6558.31), for Fields, with 2 df (= 374.29), and for "SF," with 30 df (= 4674.38). The quotation marks indicate sources not identified correctly because of our having ignored the grouping factor. In fact, the source identified as "Students" reflects not only student differences but also classroom differences.

Next, analyze each group separately. The separation will allow us to exclude the group effect. For Group 1, Ms. Brooks's class, we get SSs for "Students," with 8 df (= 4982.96), for "Fields," with 2 df (= 40.52), and for "SF," with 16 df (= 2645.48). For Group 2, Mr. Novak's class, we get "Students," with 6 df (= 1314), "Fields," with 2 df (= 1093.24), and "SF," with 12 df (= 1269.43). Then pool the corresponding sources from these group results. "Students" has 14 df and its SS is 6296.96; "SF" has 28 df and its SS is 3914.91. We don't need the "Fields" source from this analysis, as it was not affected by ignoring classrooms in the first ANOVA.

The pooled "Students" source does not include classroom effects because the variability within each classroom was extracted separately. The confounding introduced by the nesting relationship means this source actually incorporates its interaction with the source under which it is nested. That interaction, Students * Classrooms, would be the normal error term for Classrooms and so the pooled "Students" source furnishes the error term for classrooms here. The classroom effect can be isolated by subtracting the pooled "Students" source from the "Students" source obtained earlier when the grouping factor was ignored. So the SS for classrooms is 261.35 (6,558.31 − 6,296.96), with 1 df.

Similarly, the pooled "SF" is free of classroom effects. This source would be the normal error term for fields. However, the familiar impact of nesting is that this source is confounded with the three-way interaction and so serves as a shared error term. The classroom by fields interaction is determined by subtracting this error term from the "SF" term obtained from the first ANOVA, in which classroom effects were ignored. Thus, the CF SS is 759.47 (4,674.38 − 3,914.91), with 2 df. The final table has the same format, with separate error terms, as it would if there had been equal group sizes. Disappointingly, perhaps, the computers had no impact.

Source	df	SS	MS	F
Classrooms	1	261.35	261.35	<1
Error	14	6296.96	449.78	
Fields	2	374.29	187.15	1.34
CF	2	759.47	379.74	2.72
Error	28	3914.91	139.82	

This laborious method could be used to analyze the designs with equal group sizes discussed in chapter 10, although of course there is no practical reason to do so. The point is that because proportionality includes equality as a special case, any analysis that works for proportional group sizes must also work for equal group sizes.

Zero Implantation

A special case of systematic empty cells arises when participants permanently withdraw from a study. Structurally, we are considering a design in which subjects are nested under treatments and crossed with the repeated-measures factor, time periods. This situation might arise if patients in a therapeutic program that extends over time elect to drop out of the regimen or if a stressful condition induces illness or even death among participants. In such cases, the customary assumption of haphazardly occurring missing scores is clearly wrong (Weiss, 1991). Observations that occur late in the project are more likely to be missed, and once an empty cell has occurred for a given subject, the likelihood that cells for subsequent time periods for that subject will also be empty is high.

The inequality in rates of withdrawal across the various experimental groups is an important component of the evaluation of the grouping factor. A proper analysis of attrition takes into account the finality of withdrawal on the estimate of the treatment variable as well as on the estimate of the temporal effect. In many cases of behavioral interest, postwithdrawal performance on the measure examined in the study will differ radically from that observed prior to withdrawal. In such cases, extrapolating the missing scores from data actually collected may lead to absurd conclusions about treatment efficacy.

A general solution to this problem requires a theory of attrition that predicts post-withdrawal scores. I have proposed (Weiss, 1987) a method that is applicable in one specific (and perhaps not uncommon) circumstance, namely when performance of the relevant behavior stops after the participant leaves the study. The method is known as zero implantation, because it calls for the use of zeros as estimates of the missing scores. Zero implantation is a specific kind of data imputation, one based on behavioral theory rather than on statistical reasoning, but that terminology was not known to me when I published the method.

The analysis was first developed to deal with patients dropping out of an experiment designed to manipulate compliance with pill-taking instructions. Here the measure of performance was the number of pills taken properly, a measure that has obvious validity in terms of medical benefit. Patients who withdrew before the scheduled end of the project did not derive maximal value from the medicine. Just as differences on the performance measure tell of differential effectiveness of the particular compliance intervention distinguishing one group from another, so do differential attrition rates. The basic presumption underlying the method is that zero is the appropriate number to characterize the number of pills taken after withdrawal. The rationale is that if the patients wanted to keep taking pills, they would remain in treatment; pragmatically, also, patients generally have no access to pills after dropping out.

The key substantive assumption underlying the method is that zero is the proper estimate of postwithdrawal performance. A corresponding technical assumption is also required, one that concerns the scale property of the measure employed in the study. Only a ratio scale can be used because the use of zero as an estimate requires that the number zero correspond to the worst possible performance. This limitation rules out behavioral indices that may be positive or negative, such as pounds lost or blood-pressure changes. Measures that accumulate the occurrences of a directed behavior will have the desired ratio property. If these behaviors are unlikely to occur outside of the experimental setting, then zero implantation will be sensible. For example, an educational researcher might compare programs designed to help marginal students by looking at the progress points (GPA · units) earned each semester. In terms of evaluating an intervention, the student's leaving the institution is a catastrophic event and as such should have a marked negative impact on the intervention associated with it.

Because zeros replace all of the missing scores for a subject, early withdrawal will have enormous impact. The earlier the withdrawal, the lower will be the mean for that subject and thus the lower will be the group mean. Zero implantation may be viewed as a way to integrate observed scores and attrition to achieve an overall measure of treatment effectiveness.

The statistical price for the implantation is minimal. Implanted zeros are not free to vary as real scores do, and so one must remove one *df* for each estimated score. The adjustment is made on the error term used to test the sources that involve time periods. The resulting total *df* will be one less than the number of actual scores collected.

The technique may be illustrated with fictitious data from a medication compliance study in which the dependent variable is the percentage of pills properly taken by each patient during the previous month.[3] Each patient was randomly assigned to one of two intervention groups. Six measurements were scheduled for each patient, but some of the volunteers withdrew from treatment. The goal of the analysis is to ascertain the differential efficacy of the two treatments. An x denotes a missing score.

	Treatment Group 1							Treatment Group 2					
	Month							Month					
	1	2	3	4	5	6		1	2	3	4	5	6
Patient 1	83	72	91	91	76	83	Patient 6	87	83	86	77	81	x
Patient 2	57	84	52	63	74	82	Patient 7	81	x	x	x	x	x
Patient 3	96	80	33	52	68	66	Patient 8	78	87	84	x	x	x
Patient 4	85	91	73	42	x	x	Patient 9	87	x	x	x	x	x
Patient 5	87	59	65	72	83	75	Patient 10	82	89	x	x	x	x

The analysis follows the procedure given in chapter 11 for nested designs, with zeros replacing the missing scores. When an implanted score, such as a zero in this example, is entered, double-click on the value after typing it. That will change the color of the score to red (in the grayscale rendition below, you'll have to imagine the zeroes are red), which serves as a signal to FACTORIAL ANOVA that a *df* adjustment involving that score is called for. A second double-click turns the score back to black and removes the adjustment.

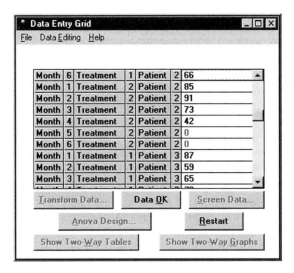

Note the *df* adjustment in the second error term in this table:

Source	df	SS	MS	F
Groups	1	17784.8	17784.8	8.65*
Error	8	16444.8	2055.6	
Months	5	18755.1	3751.0	2.94
GM	5	6015.1	1203.0	<1
Error	20	25544.4	1277.2	

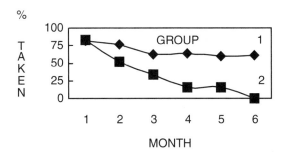

If we consider only the data actually collected, ignoring the attrition, patients in treatment group 2 on the average took a higher percentage of pills (83.5%) than those in treatment group 1 (72.7%). However, the zero-implantation analysis radically alters our perspective on the results, correctly in my view. The graph shows that the mean percentages taken were higher in treatment group 1, reflecting the fact that those patients persisted with the treatment. The method assumes that dropouts don't take pills.

Snapshot Analysis

In applied research, it is often the case that some participants are farther along in the course of the treatment than others. In clinical settings, for example, patients eligible for the study will be randomly assigned to experimental groups as they come along. If measures are taken periodically in such a situation, then at a given moment some participants will have produced more scores than others. By the end of the study, if all goes according to plan, these discrepancies will have been resolved. The plan may call for each patient to be treated for six months with induction carried out over a year; thus completion will require eighteen months. But it may be of practical interest to assess the grouping variable prior to the scheduled end of the project. The intermediate results might allow the researcher to abandon an unpromising line of research. The desired SNAPSHOT view of the data is complicated by the unequal number of scores per subject.

A simple solution to this problem may be obtained when the experimental design is balanced (Weiss, 1985a). The requirement of balance means that at the time of the snapshot, there must be in each group the same number of subjects who have produced a given number of scores. This can be achieved by using random permutations for the assignment of subjects to groups; at worst there will be a few extra scores at the time of the analysis. One may either ignore these or wait for balance.

Balance allows a proportional cell size structure in a repeated-measures design in which participants are nested under treatments. This produces an orthogonal analysis. The usual rules for generating the error terms apply.

The example shows the appropriate balance:

Time	Treatment 1				Treatment 2				Treatment 3			
	P_1	P_6	P_7	P_{12}	P_2	P_4	P_9	P_{10}	P_3	P_5	P_8	P_{11}
1	7	6	5	2	6	3	4	7	5	2	3	4
2	5	8	9	8	4	3	5	5	8	2	5	2
3	4	4	7		6	3	5		4	3	4	
4	6	7			5	4			5	3		
5	9	3			5	3			6	5		
6	8				3				4			
	39	28	21	10	29	16	14	12	32	15	12	6

In this study, participants were assigned to the treatments in sets of three (using random permutations). There were four such sets at times one and two, three at time three, and so on, until there was only one set available at time six. The subscripts used for participants reflect their order of induction into the study. The score for each participant is the number of miles run during the previous month. The question of interest is whether the experimental treatments induce differential amounts of reported exercise.

The nested source, participants, is (temporarily) presumed to have four levels and to be crossed with treatments (so that for the $SS_{Participants}$ computation the actual P_1, P_2, and P_3 are regarded as P_1, the actual P_6, P_4 and P_5 are regarded as P_2, etc.). The sources that can be extracted from this design are $SS_{Participants}$, $SS_{Treatments}$, SS_{PT}, and $SS_{Residual}$.

$$SS_{Participants} = P_1^2/18 + P_2^2/15 + P_3^2/9 + P_4^2/6 - T^2/N$$
$$= 100^2/18 + 59^2/15 + 47^2/9 + 28^2/6 - 1{,}140.75$$
$$= 22.98$$
$$SS_{Treatments} = (T_1^2 + T_2^2 + T_3^2)/16 - T^2/N$$
$$= (98^2 + 71^2 + 65^2)/16 - 1{,}140.75$$
$$= 38.625$$
$$SS_{PT} = P_1^2/6 + P_6^2/5 + P_7^2/3 + P_{12}^2/2 + P_2^2/6 + \cdots + P_{11}^2/2$$
$$- SS_T - SS_P - T^2/N$$
$$= 39^2/6 + 28^2/5 + 21^2/3 + 10^2/2 + 29^2/6 + \cdots$$
$$+ 6^2/2 - 38.625 - 22.98 - 1{,}140.75$$
$$= 15.31$$
$$SS_{Residual} = \Sigma X^2 - T^2/N - SS_T - SS_P - SS_{TP}$$
$$= 1{,}310 - 1{,}140.75 - 38.625 - 22.98 - 15.31$$
$$= 92.33$$

The standard scheme for nested designs is used to determine the error term against which $MS_{Treatments}$ *will be tested.* $SS_{Participants}$ *is pooled with* SS_{PT} *to yield* MS_{Error}. *Here* $F_{2,9} = 19.31/4.25 = 4.54^*$.

The term labeled "Residual" includes the variation over time periods and the interactions involving time periods. Because this term is not meaningful with differing numbers of periods for the various participants, the *df* (36 in the example) are wasted in this analysis. This statistical inefficiency is the price for the snapshot analysis. Only at the completion of the experiment will time periods have the status of a proper factor in an ordinary repeated-measures nested design.

The snapshot algorithm has been incorporated into the program SNAPSHOT (Elder & Weiss, 1987), whose opening screen is shown below. The program goes beyond the text[4] to allow for two or three experimental factors.

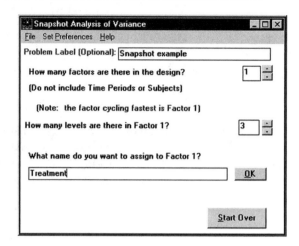

Exercises

12-1. An anthropologist evaluated the effects of diet and religious practice on growth in a natural experiment. The heights of adult males in four small tribes, all of whose members were from the same ethnic stock, were measured. Analyze the data.

Tribe A: Eats meat, prays daily for strength
Heights of adult males: 64, 62, 68, 65, 71, 69, 68, 68
Tribe B: Eats meat, atheist
Heights of adult males: 70, 66, 68, 65, 64
Tribe C: Vegetarian, prays daily for strength
Heights of adult males: 59, 66, 64, 66, 62, 61
Tribe D: Vegetarian, atheist
Heights of adult males: 64, 62, 58, 63, 64, 68

12-2. Eight groups of 10 volunteers were recruited for an experiment on memory. Each subject saw one hundred words, presented visually using a slide projector, during the first session and another hundred words during the second session. Fifty of the second hundred were repeats, and fifty were new. The subject's task was to identify which words were old and which were new, and the score was the number correctly labeled during the second session. The eight groups were constructed according to a 2 × 2 × 2 design with the factors corresponding to characteristics of the slides. Half of the subjects saw nouns and half saw verbs, half of the subjects got words of four letters and half got words of six letters, and half of the subjects saw each slide for 5 seconds, while the others saw each slide for 15 seconds. Unfortunately, not all of the subjects showed up for the second session. Analyze the scores.

MISSING DATA 179

	Nouns				Verbs			
	4 letters		6 letters		4 letters		6 letters	
	5 seconds	15 seconds	5 seconds	15 seconds	5 seconds	15 seconds	5 seconds	15 seconds
	76	36	43	37	94	74	67	60
	43	45	75	22	80	64	64	54
	65	47	66	22	81	68	70	51
	42	23	56	25	80	72	65	49
	60	43	62	11	80	62	60	38
	78	43	51	27	69	78	55	55
	66	54	63	23	80	61	57	56
		45	52	25	58		66	58
		41	50	31			79	
		40					80	

12-3. The day after Christmas, six of the fathers in my apartment building happened to meet at the laundry, and they compared notes on how long it had taken to assemble some heavily advertised toys. Not surprisingly, Santa brought the same toys to almost every home. Analyze the data to see if there are differences in assembly time (in minutes) for four popular toys.

Fathers	Racer 500	Guillotine	Batto	Forest Fire
Joe	95	42	18	140
Bill	72	x	10	100
John	53	32	7	89
Dave	86	45	20	115
Ken	x	39	15	115
Steve	63	51	24	27

12-4. Twenty-four sophomores were hired to listen to a rock album and then rate its quality on a scale of 0 to 100. Listening alone, each student heard either *The Robots Live* or *The Bouncing Pebbles*. To simulate actual concert conditions, half of the students drank a half-liter of wine while listening, and the other half smoked a special import. Here are the ratings; not all of the listeners filled out the rating form. Compare both the albums and the accompaniments.

Wine		Smoke	
Robots	Pebbles	Robots	Pebbles
40	71	63	83
13	82	48	95
61	45	79	89
52	64	65	
71	77		
53	95		

12-5. Some of the female students in my statistics class complained that the course was biased in favor of men because men are better at math. One of the men responded that the women had been acculturated to be fearful of numbers, and he thought this fear could be trained out at the Psychology Department Anti-Anxiety Clinic. We decided to conduct an experiment on the class members to test the efficacy of the clinic. Half of the 10 male students and half of the 20 female students were chosen (with a random number table) and assigned to go to three sessions of the clinic. Then the scores on the final exam were analyzed to test the questions of interest: is the training effective and is this effectiveness accentuated for women? Analyze the data.

	Without clinic	With clinic
Men	3	79
	64	82
	91	83
	47	52
	78	65
Women	63	79
	84	85
	71	91
	53	84
	42	62
	67	53
	95	88
	77	77
	32	93
	62	99

12-6. A health psychologist designed a program to increase the patient's desire to exercise. The control group was shown videos of Britney Spears exercising, and the experimental group was shown an exciting multimedia presentation stressing the short-term and long-term benefits of regular exercise. The patients were randomly assigned to one of the two groups and then were secretly monitored during their weekly sit-up workout at the Cal State Health Club. Every week, the patient watched the presentation, did sit-ups, and then relaxed in the spa. The program was scheduled to last for eight weeks, and the score was the number of sit-ups performed during the session. An x means that the patient did not attend the session. Assess the effectiveness of the programs.

	Experimental Weeks									Control Weeks							
	1	2	3	4	5	6	7	8		1	2	3	4	5	6	7	8
Patient A	42	37	22	63	54	38	62	75	Patient H	97	85	73	52	x	x	x	x
Patient B	48	51	15	48	62	39	47	38	Patient I	74	64	49	73	38	x	x	x
Patient C	36	29	38	53	47	39	52	41	Patient J	55	43	65	58	72	64	57	68
Patient D	25	27	32	43	28	32	25	x	Patient K	29	53	x	x	x	x	x	x
Patient E	38	46	52	34	39	41	47	48	Patient L	72	83	75	67	54	38	x	x
Patient F	28	31	36	25	39	42	x	x	Patient M	48	51	65	49	58	x	x	x
Patient G	26	42	53	39	x	x	x	x	Patient N	39	43	54	41	33	x	x	x

12-7. In an attempt to help young faculty members succeed in their new jobs, the university instituted a mentoring system in which senior faculty were assigned to help their junior colleagues publish in professional journals. While the junior faculty members who published sufficiently were to be given the customary reward of a tenured position, the mentors were promised a cash bonus for each article published by their mentees. Because the system was expected to be costly to the university (just as hiring replacements for lost junior faculty members is costly), it was deemed advisable to evaluate the new program via an experiment. As new faculty members were hired, they were randomly assigned to one of three groups. Group 1 was a traditional group, with no mentoring. Group 2 had mentors assigned, but no cash bonus was offered to the mentors. Group 3 incorporated the monetary incentive for mentors. The annual survey of publications provided the data. The study was carried on for five years; new faculty members were added each year. Evaluate the mentoring program.

Professor	Year Hired	Group	Publications in 1984	1985	1986	1987	1988
DP	1984	1	0	1	0	1	1
KM	1984	1	1	2	0	1	1
LZ	1984	2	2	1	0	0	0
PR	1984	2	1	0	0	0	0
JH	1984	3	1	1	1	2	0
MK	1984	3	1	2	3	2	2
BD	1985	1		0	1	1	0
AG	1985	2		1	0	0	0
FV	1985	3		1	3	3	2
GR	1986	1			1	1	2
NA	1986	1			0	1	0
GM	1986	1			1	1	0
JS	1986	2			2	1	2
DR	1986	2			1	3	2
LN	1986	2			0	1	0
WB	1986	3			1	2	4
CT	1986	3			3	3	2
RP	1986	3			2	1	3
JD	1987	1				1	1
ST	1987	1				1	2
LS	1987	2				1	1
NC	1987	2				3	3
BW	1987	3				2	3
SQ	1987	3				3	3
HD	1988	1					0
JL	1988	2					0
FN	1988	3					2

12-8. When the NCAA threatened to shut down the university's baseball and basketball programs because an athlete had committed suicide, the administration took immediate action. The two teams' starters were enrolled in separate group therapy programs led by psychology graduate students. The therapy emphasized nonathletic components of self-esteem. Each athlete was given a depression inventory (0 = not at all depressed, 50 = very depressed) at the beginning of the year prior to therapy and then was tested again at the beginning of each quarter. The main concern was whether the athletes' scores changed over the course of the year. Dwayne, a starting guard on the basketball team, missed the spring testing session when he had to go home for a family emergency.

Baseball Team	Fall	Winter	Spring
Bob, 1B	45	38	20
Pete, 2B	27	22	25
Bill, 3B	40	20	17
George, SS	10	13	8
Darryl, RF	48	22	28
Sam, CF	35	25	18
Jose, LF	37	28	31
Brent, C	47	43	29
Ron, P	38	35	37

Basketball Team	Fall	Winter	Spring
Ed, G	20	17	12
Dwayne, G	28	22	x
Joe, F	41	36	28
Patrick, F	7	6	9
Keith, C	27	25	21

12-9. In response to the NIH's solicitation for research projects aimed at early diagnosis and treatment of AIDS, a research team proposed to investigate the joint effectiveness of a new drug and a behavioral intervention. As soon as diagnosis occurred, the patient was asked to enroll in the study, and after accepting, the patient was assigned to one of four groups. Group 1 received the standard treatment, AZT, and was considered as control group. Group 2 received the new drug, referred to as XYZ until its efficacy was established. Patients in Group 3 were given AZT along with instruction in meditation to help manage stress; they were also taught to visualize their immune system repelling invaders. Group 4 patients received XYZ and the same instructional program as those in Group 3. Treatment effectiveness was measured via monthly blood samples; the dependent variable was the number of T-cells per cubic millimeter of blood. The grant provided funding for one year, at the end of which a decision would be made on whether renewal was in order. While it would be ideal to track the time course of treatment efficacy, in practice the small number of available patients at a given time meant that a design dedicated to that issue would require several years. Therefore, the researchers followed customary FDA procedures in attempting to show that a proposed treatment is more effective than an established one, rather than attempting to demonstrate effectiveness in an absolute sense. Given the results, are XYZ and meditation worthy of further investigation?

	Group 1						Group 2					
	AK	BZ	RV	GT	LS	WT	MW	LL	CD	HR	MN	DR
Jan	410						413					
Feb	426	385					511	375				
Mar	416	393					518	386				
Apr	425	408	395				524	392	387			
May	418	422	402				529	408	395			
Jun	427	438	401				527	415	402			
Jul	434	429	416	388			536	426	454	398		
Aug	435	427	456	391	386		552	430	447	411	395	
Sep	445	439	462	401	392		564	451	462	412	399	
Oct	449	451	464	415	399	404	569	462	475	415	408	386
Nov	462	468	462	426	411	414	584	473	485	433	416	399
Dec	471	483	475	431	433	426	595	486	498	453	421	413

	Group 3						Group 4					
	DF	AW	TR	JT	BE	GN	KJ	MJ	NT	AY	PR	HL
Jan	396						401					
Feb	393	420					438	387				
Mar	406	429					476	399				
Apr	415	437	386				515	428	382			
May	424	489	396				534	443	399			
Jun	429	503	408				567	462	426			
Jul	441	511	420	375			578	473	434	380		
Aug	451	507	432	396	381		593	487	441	399	372	
Sep	463	516	442	404	387		604	499	463	418	386	
Oct	473	520	460	411	396	391	621	515	479	435	407	388
Nov	491	525	474	427	409	415	636	528	496	448	419	418
Dec	511	529	490	442	416	423	652	546	521	457	433	436

Answers to Exercises

12-1.

Source	df	SS (raw)	MS	F
Prayers	1	0.002	0.02	<1
Diet	1	13.354	81.14	9.63*
PD	1	0.048	0.29	<1
Within cells	21	176.908	8.42	

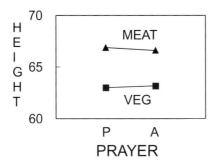

12-2.

Source	df	SS (raw)	MS	F
Word type	1	792.85	6621.31	79.92*
Number of letters	1	288.59	2410.06	29.09*
Duration	1	712.58	5950.99	71.83*
WN	1	5.21	43.54	<1
WD	1	108.85	909.07	10.97*
ND	1	37.86	316.16	3.82*
WND	1	9.45	78.93	<1
Within cells	60	4970.78	82.85	

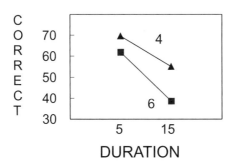

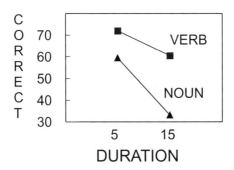

12-3.

Source	df	SS	MS	F
Fathers	5	3041.5		
Toys	3	22984.5	7661.5	16.27*
Error (FT)	13	6120.3	470.8	

Estimated scores: Bill—Guillotine = 37.65; Ken—Racer = 61.35.

Note: These estimated scores were based upon the Racer 500 and Guillotine values for the surrounding fathers. As these choices are somewhat arbitrary, estimates based on other scores should be considered equally correct. Naturally, the ANOVA for other estimates will be (slightly) different.

12-4. This analysis should not be carried out. All of the missing scores have come from the smoke conditions; perhaps something about this treatment makes subjects less likely to carry out the rating task. The test for equality of attrition yields a significant F ratio for accompaniments (1, 20 df) of 7.35. The situation does not accord with the requirement that missing scores occur haphazardly without regard to the treatment. Further laboratory work (or perhaps a new statistical procedure—do you have a theory that generates predicted values to replace the missing scores?) would seem to be in order.

12-5.

Source	df	SS	MS	F
Gender	1	476.02	476.02	1.19
Clinic	1	1968.30	1968.30	4.92*
GC	1	1.35	1.35	<1
Within cells	26	10405.30	400.20	

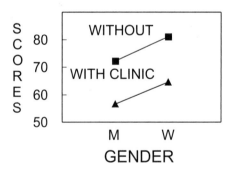

12-6.

Source	df	SS	MS	F
Groups	1	18.08	18.08	<1
Error	12	18346.79	1528.90	
Weeks	7	15847.78	2263.97	4.93*
WG	7	9925.13	1417.88	3.09*
Error	56	25703.21	458.99	

Note: In proposing the appropriateness of the zero-implantation method for this situation, I am perhaps imposing my own view about exercise on the statistical solution. My guess, doubtless an arguable one, is that patients whose condition calls for placement in an exercise program would not do situps on their own. Ideally, I could cite empirical evidence to support this theory of the behavior of dropouts; the evidence would thereby support the statistical decision. While zero implantation may be controversial in this situation, it should be kept in mind that any alternative statistical procedure also requires justification, ideally empirically based.

12-7.

Source	df	SS	MS	F
Groups	2	28.50	14.25	10.09*
(Residual)	57	29.37		
Error	24	33.88	1.41	

12-8.

Source	df	SS	MS	F
Teams	1	631.07	631.07	2.46
Error	12	3082.67	256.89	
Quarters	2	835.32	417.66	13.28*
TQ	2	64.12	32.06	1.02
Error	23	723.45	31.45	

Note: The estimated score for Dwayne's missed session was 16.32, based on Ed's and Joe's values. One *df* was subtracted from the error term that includes the subject by treatment interaction, namely the shared error.

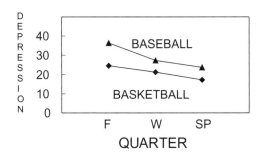

12-9.

Source	df	SS	MS	F
Drug	1	37820.89	37820.89	2.58
Meditation	1	10320.02	10320.02	<1
DM	1	34.78	34.78	<1
(Residual)	160	246342.90		
Error	20	293646.70	14682.34	

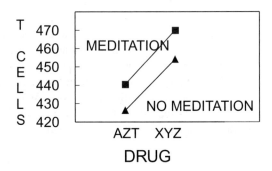

Notes

1. Strictly speaking, the unweighted means analysis given here is an approximate one in that it proceeds as though all of the means had equal variances. More complex methods, each based on different assumptions about the precise statistical hypotheses being tested, are discussed by Herr and Gabelein (1978). The various methods yield different solutions for the sums of squares. At present, there does not seem to be consensus among statisticians on a single best approach, although a least squares method is used as the default option by several of the leading commercial statistics packages. None of the standard techniques for nonorthogonal ANOVA is robust to violations of homogeneity of variance or normality (Milligan, Wong, & Thompson, 1987). When these violations are likely to arise, extra effort to achieve equal cell sizes is especially worthwhile.

2. The software is available through Pennsylvania State University's Methodology Center, whose home page is http://methodology.psu.edu.

3. The tables are from "Dropout in patient compliance studies: A suggested analytic procedure" by David J. Weiss, published in 1987 in the *Journal of Compliance in Health Care*, *2*, 73–83. Copyright 1987 by Springer Publishing Company, Inc., New York. Used by permission.

4. The SNAPSHOT program and the multifactor extension were conceived (and published!) by Elder while he was a (mere?) student in a course in which I introduced the basic scheme. I report this historical tidbit to support two of my beliefs about ANOVA: first, the field is not a closed one—there are many interesting situations yet to be explored; second, general principles are sufficient to derive advanced solutions. One need not be a professional statistician to contribute to this progress. There is a publication market for such programs, if one is inclined to such efforts.

13

Confounded Designs

In our discussion of nested designs, we encountered the concept of a design that was not fully crossed. We saw that because of the lack of crossing, certain interactions could not be estimated. In this chapter, the sacrifice of interactions is extended. Experimental designs that are far from complete factorials are introduced.

For example, consider the following trio of treatment combinations: A_1B_1, A_2B_1, A_2B_2. Here we have two factors, each with two levels, yet there are only three combinations rather than the expected four. The effect of factor A can be measured by comparing the responses to the first two combinations, while the effect of factor B can similarly be assessed by comparing the latter pair. Three cells yield 2 treatment df, and so we can ask two separate questions of the data. Normally the treatment combination A_1B_2 would also be included in a two-factor design, and then the AB interaction would be estimable as well. Without that fourth cell, we cannot estimate the interaction, but our estimates of the main effects are sensible.

Cutting one cell from a design hardly seems useful; but if we extend the reasoning, practical incomplete designs soon emerge. Let us look at a subset of a three-factor design. Suppose we present the treatment combinations $A_1B_1C_2$, $A_1B_2C_1$, $A_2B_1C_1$, and $A_2B_2C_2$. With these four combinations, all three main effects can be assessed. Factor A's contribution is given by the difference between the sum of the first two cells as opposed to the sum of the latter two. Factor B is measured by subtracting the first cell plus the third from the second cell plus the fourth. Can you see how to measure factor C? The rule is simply to combine all of the cells that have level 1 of the factor you're after and subtract the sum from the sum for the cells that have level 2 of that factor.

But now here comes the rub. How could you measure the AB interaction in this subset? The normal procedure for estimating the interaction is to compare the difference between A_2B_2 and A_2B_1 to the difference between A_1B_2 and A_1B_1. In this three factor design, then, you should compute $(A_2B_2C_2 - A_2B_1C_1) - (A_1B_2C_1 - A_1B_1C_2)$. Unfortunately, however, a little algebra shows that this way of putting the four terms together is the same as the way that yielded the effect of factor C! Since the computations for C are the same as those for AB, these sources cannot be distinguished in this design. They are said to be confounded. In a similar way it may be determined that A and BC are confounded, as are B and AC. The ABC effect cannot be measured at all, just as the two-way interaction could not be measured in our previous example with only three treatment combinations. These limitations should have struck you as inevitable, since with four cells only three *df* are available for treatment effects.

Don't lose sight of the profit, though. In our example, we were able to estimate three main effects (albeit confounded effects) with only four treatment combinations. Although the word confounded has an evil ring to it, and perhaps even calls to mind poor control over experimental variables, in its present usage it is strictly a technical term. Incomplete designs yield less information than complete factorials, but fewer treatment combinations need to be presented. This tradeoff merits detailed examination, especially in a world in which financial constraints may play a role in the design of experiments.

Intentional confounding of sources is the central theme. Confounding means that a single sum of squares contains two or more conceptually distinct sources. Because the logically distinct sources are pooled, they cannot be estimated separately. Further confusion can arise because the pooled sources may add, thus augmenting the sum of squares, or they may cancel one another's effects, thus leaving a negligible sum of squares.

Obviously, the researcher is not aiming at chaos. The usual assumption when a confounded design is employed is that only one of the sources in a pool has an effect; any others are said to contribute essentially nothing. This assumption should be empirically based; but in practice, it is often based primarily on theoretical considerations.

Confounded designs are incomplete factorials, constructed in such a way that specified sources become merged. Which sources become merged depends on the nature of the design; this matter can be complicated. Usually the researcher selects a design in which main effects are confounded with specific interactions that are expected to be insignificant.

The point of confounding is to save experimental labor. Of course, the saving is not of effort (researchers are never lazy) but of money or time. Sometimes it is simply impractical to conduct a complete factorial design. In addition, many of the observations in a complete factorial seem poorly rewarded. For example, consider a 2^6 (that is, $2 \times 2 \times 2 \times 2 \times 2 \times 2$) design in which the researcher is interested primarily in the six main effects and secondarily in some of the two-way interactions. The 64 cells in the design generate 63 *df*, and perhaps only 12 of the *df* are used for tests of interest to the researcher. The *df* used to test the three-, four-, five-, and six-way interactions are essentially wasted. Using the observations to generate degrees of freedom applied only to interesting sources seems more efficient. If only 12 *df* are needed for the effects of interest, then either the other *df* should contribute to the error term or fewer observations should be collected, but collecting scores to test uninteresting effects seems to get the worst of both worlds.

The danger inherent in confounding should not be minimized. Recall that main effects are often uninterpretable in the presence of interaction. Yet confounded designs give up on the possibility of estimating many interactions. The researcher must be sure that any interactions that would cloud the interpretation of important main effects are indeed nonsignificant.

Latin Squares

Perhaps the most popular of confounded designs is the Latin square. Possessed of great appeal to lovers of symmetry, the Latin square is an arrangement of symbols (originally Roman numerals—hence the name) so that each appears once in each row and column. For example, a 3 × 3 square might be:

I	II	III
II	III	I
III	I	II

The symbols may be rearranged while still preserving the Latin square property. The number of different possible arrangements increases dramatically as the size of the square increases. The square may be used to superimpose an additional factor onto a two-factor design with the same number of levels on both factors. The resulting design has three factors, but they are not crossed in the usual manner. The representation of this design employs the symbols from the previous square as subscripts of the C factor.

	A_1	A_2	A_3
B_1	C_1	C_2	C_3
B_2	C_2	C_3	C_1
B_3	C_3	C_1	C_2

In the Latin square, the factors are not fully crossed with one another (notice there is no $A_1B_1C_2$ or $A_1B_1C_3$), but they are crossed on a pairwise basis. Each level of each factor is paired with each level of the other two factors considered separately. This can be seen by rewriting the previous design so that factor C is on the outside.

	C_1	C_2	C_3
B_1	A_1	A_2	A_3
B_2	A_3	A_1	A_2
B_3	A_2	A_3	A_1

The rewriting to illustrate this point could also have been accomplished by putting factor B on the inside. The Latin square design is perfectly symmetrical.

The square tells the researcher which stimulus combinations to present. The corresponding statistical analysis yields fewer sources than the usual three-factor design, because interactions are not estimable. Each main effect is completely confounded with the interaction of the other two factors. This confounding is the price of the savings afforded by the Latin square; its consequence is that the design can responsibly be used only when all interactions are known a priori to be negligible.

We can illustrate the computational procedure with an example involving a 4×4 Latin square. The university cafeteria offered sixteen students free lunches in an attempt to learn ways to make their meals more attractive. Each student ate lunch and then rated the satisfaction gained from the dining experience. Each meal was served in one of the four campus facilities: "The Frosh Firepit" (A_1), "The Loser's Lanai" (A_2), "The Burrito Bowl" (A_3), and "The Den of Iniquity" (A_4). Four different campus specialties were served: macaroni and cheese (B_1), beef stew (B_2), chop suey (B_3), and meat loaf (B_4). The innovative feature of the research was the use of background music to accompany the meal. Three albums were used, Hotel California (C_1), Meat Loaf (C_2), and Joy of Cooking (C_3), along with no music (C_4). The treatment combinations used, as well as the ratings (on a 100-point scale), are given in the square.

	A_1	A_2	A_3	A_4
B_1	C_2 33	C_3 41	C_1 28	C_4 46
B_2	C_4 61	C_1 39	C_3 28	C_2 42
B_3	C_1 51	C_4 55	C_2 39	C_3 27
B_4	C_3 67	C_2 59	C_4 72	C_1 49

The analysis is carried out in two stages. Each stage is an abbreviated analysis of a two-factor design. To begin, work with the two outside factors (here A and B), ignoring the interior one. Compute ΣX^2 (= 36,891) and T^2/N (= 33,948.06). Then find the main effect sums of squares from the marginal totals (SS_A = 34,341.25 − T^2/N = 393.19; SS_B = 35,349.25 − T^2/N = 1,401.19). Next, rewrite the square so that either of the two exterior factors goes inside. Repeat the two-way analysis in order to extract the missing main effect (SS_C = 34,785.75 − T^2/N = 837.69). The other terms in this second analysis may be used as an arithmetic check. If you prefer, these two two-way analyses may be carried out by the computer (but you still need to rewrite the square).

One new term needs to be computed. Not all of the available df are used for the main effects. Left over are $(N - 1) - 3 \cdot (J - 1)$ df, where J is the number of levels on each factor. The sum of squares associated with these degrees of freedom is known as the Latin square error or the remainder. It consists of portions of the interactions that are not confounded with the main effects. Because the user of a Latin square is willing to assume that interactions are really zero, then if no better estimate of error is available this Latin square error may be used as the error term for F tests. A simpler computational expression for the df for the Latin square error is $(J - 2) \cdot (J - 1)$. The sum of squares for the Latin square error is determined by subtraction. This remainder is given by: $\Sigma X^2 - T^2/N - SS_A - SS_B - SS_C$. If the computer is producing the table, then the formula is $SS_{Total} - SS_A - SS_B - SS_C$. Now we can write the ANOVA table for our example:

Source	df	SS	MS	F
A (facility)	3	393.19	131.06	2.53
B (meal)	3	1401.19	467.06	9.01*
C (music)	3	837.69	279.23	5.39*
L.S. error	6	310.87	51.81	

Greco-Latin Squares

The confounding featured in the Latin square can be extended still further. For many Latin squares, an additional Latin square can be superimposed using, say, Greek letters, in such a way that each combination of symbols occurs exactly once within the square. For example:

	A_1	A_2	A_3
B_1	I α	II β	III Γ
B_2	II Γ	III α	I β
B_3	III β	I Γ	II α

Two squares that can be aligned in this fashion are said to be orthogonal. Not all Latin squares can be; for example, there are no orthogonal 6 × 6 squares.

Both sets of symbols can be used to embed factors into a two-factor design. The resulting four-factor design is analyzed in two stages, with the first analyzing the AB design and the second requiring rewriting of the square so that C and D are brought outside.

In the Greco-Latin square, each variable is confounded with a pair of interactions. Despite the risks inherent in heavy confounding, the Greco-Latin square has occasionally proven useful in research, particularly when variables such as temporal order or physical placement are involved. The researcher wants to incorporate such control factors into the experiment, but does not want to devote large percentages of the research energy to exploration of perhaps uninteresting interactions.

There is an important sense in which confounded designs are very efficient. If a researcher has several variables whose impact must be assessed, and has limited experimental resources, then a confounded design may be optimal. The reason for this assessment is that a minimal number of observations is used to get the estimates of the main effects. Consider a 3 × 3 Greco-Latin square, a case in which there are four effects to be investigated. Each effect uses up two *df*, which means that every one of the *df* is being employed to answer the questions of central concern to the researcher. No *df* are expended on estimating interactions. For preliminary investigations especially, Latin square designs can be useful to determine quickly whether particular treatments have any effect at all.

Danger and Inconveniences in Latin Squares

Having no or few tests of interactions is a mixed blessing. The validity of statements about main effects depends upon there being no interactions to cloud the interpretation. Thus, the researcher is critically dependent upon the prior assumption that interaction is nonexistent. It is all too easy for an investigator to employ a Latin square mechanically, without securing the necessary background information. This is a trap that should be avoided.

In a similar way, the use of the Latin square error as an estimate of random variability is hazardous. Again, the researcher's conclusions depend upon an absence of interaction. Alternative designs that preserve the spirit of the Latin square but have more reliable error terms offer superior alternatives.

Superior Error Terms

The simplest path to a better error term is through the use of replication. The same square may be used repeatedly, and the resulting within-cells mean square serves as the error. Alternatively, one may construct a repeated-measures design in which each subject receives some of the treatment combinations within one of several different orthogonal squares used in the study. Nesting also may be employed, with its confounding being untangled in accord with the customary rules.

In the repeated-measures designs, the treatment by subject interactions may be confounded with various substantive interactions. Even so, the error term or terms so composed are in general preferable to the Latin square error; at least the primary component is a standard error term.

When a superior error term is available, the term that corresponds to Latin square error is labeled "residual," and it may be tested for significance. The residual is a pooling of interactions that cannot be identified with any specific source. It is customary to place the residual term in the ANOVA table after the three main effects. The sum of squares is computed by subtracting "everything" from SS_{Total} (i.e., $\Sigma X^2 - T^2/N - SS_A - SS_B - SS_C - SS_{Error}$). From the computer-generated tables, $SS_{Residual}$ may be (manually) calculated as $SS_{Total} - SS_A - SS_B - SS_C - SS_{Within}$. If the design entails repeated measures, then instead of subtracting SS_{Within}, subtract all of the individualized error terms. Degrees of freedom are determined in an analogous fashion. If the residual proves sizable, a complete factorial design is probably in order, as the Latin square assumption about interactions now appears doubtful.

Randomizing the Square

Because only certain treatment combinations are used, it is especially important for the researcher to avoid bias in the pairings. Randomization avoids this difficulty and is a necessary step in utilizing a Latin square design in an experiment. The classical way to randomize the square is to begin with one in standard form. This means the first row and first column are in numerical order. There is more than one standard square for squares larger than 3×3. For larger squares, it has been traditional to randomly select a standard square of the desired size from the table given in chapter 4 of Cochran and Cox (1957). Next, the rows should be shuffled, followed by a separate shuffling of the columns. Finally, the levels should be randomly assigned to the indices for each factor. The modern alternative is to let a computer program such as RANDOMLS (a descendant of Strube's [1988] BASIC program) carry out the entire randomization process in a flash. The program starts with a standard square of the desired size, then a click yields a randomized square as shown below. The numbers within the square prescribe the level of the third factor that goes with each row-column pair. For example, if A denotes the row factor and B the column factor, then one of the sixteen combinations administered in accord with this randomization should be $A_2B_3C_2$.

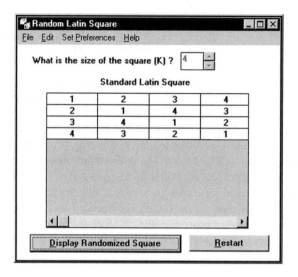

Fractional Factorial Designs

Designs in which controlled confounding is featured provide a way for researchers to conduct multifactor experiments that would otherwise be impractical. I once conducted a single-subject investigation in which there were ten factors. Even with only two levels per factor, a complete design would have required 1,024 observations from the volunteer just to get through all of the stimulus combinations. And I wanted to use a within-cells error term to get a precise measure of variability! Obviously some alternative approach was in order.

Using a fraction of a factorial design offers a solution to the problem of too large a set of treatments. A one-half replicate cuts the number of treatments in half; a one-sixteenth replicate affords an even greater reduction. The smaller the fraction, the greater the degree of the confounding. Even for extremely reduced designs, though, it is possible to estimate many desired effects up to the limits of the available df. However, each of the effects sought will be confounded, and the smaller the fraction used, the more sources with which each effect will be confounded. Constructing the experimental design so that the desired effects are estimable and not confounded with one another is an art requiring considerable expertise. These complexities will be introduced here, but for practical applications most researchers consult chapter 6A of Cochran and Cox (1957). The one caution that cannot be overemphasized when one uses a confounded design concerns the critical assumption that sources confounded with the effects of interest are held to be negligible. If the researcher cannot justify this assumption, then the confounded design is inadvisable.

Fractionating a Design

We can best illustrate the fractionating process with small designs, although its advantages are most pronounced for large ones. Consideration will be restricted to cases in which all factors have only two levels (the 2^P series). This is the most common situation in practice because two levels are sufficient to demonstrate the existence of an effect.

Recall the terminology of chapter 4. Equation 4-3 expressed the interaction effect for a two-factor design:

$$\bar{x}_{12} - \bar{x}_{11} - \bar{x}_{22} + \bar{x}_{21} \qquad (4\text{-}3)$$

Now we rewrite the equation, for convenience keeping only the indices:

$$12 + 21 \parallel -11 - 22$$

The bars separate the four cells of the 2^2 design into two sets, both of which are balanced with respect to the main effects of both factors. Each level of each factor appears once in each set. Thus any difference between the two sets is not a function of either main effect, but only of the interaction.

In the same way, a 2^3 design can be split up into two sets such that any difference between the sets is a function only of the three-way interaction. For the 2^3 design, the balancing extends beyond the main effects to the two-way interactions; all of the pairwise combinations of levels (i.e., 11, 21, 12, 22) appear once in each set.

$$112 + 121 + 211 + 222 \parallel -111 - 122 - 212 - 221$$

Suppose we were to try to estimate the effects in the design from one of these sets. If we denote the three factors as A, B, and C respectively, then we can construct a table that shows how the responses to each treatment combination combine to yield the factorial effects:

Treatment Combinations	Factorial Effects						
	A	B	C	AB	AC	BC	ABC
$A_1B_1C_2$	−	−	+	+	−	−	+
$A_1B_2C_1$	−	+	−	−	+	−	+
$A_2B_1C_1$	+	−	−	−	−	+	+
$A_2B_2C_2$	+	+	+	+	+	+	+

The way to read this table is to look at the sign given to the treatment combination as it contributes to the effect. For instance, let's look at the A main effect. If we add the four treatment totals algebraically in accord with the tabled signs, we obtain a quantity related solely to the difference between the responses to levels 1 and 2 of the A factor. The contributions of the other two factors, taken both singly and in concert, balance out over the summation. This table is an alternative way of writing equations 4-1 through 4-3. The signs correspond to the coefficients (+1, −1) in a specific comparison.

Because only half of the treatment combinations in a full 2^3 design are given, the table presents some oddities. One is that ABC cannot be estimated; all of its signs are +, a pattern that does not yield a comparison because the sum of the coefficients is not zero. The other oddity is that the effect pairs A and BC, B and AC, and C and AB have the same patterns of signs. If the other four treatment combinations were given, differences in the sign patterns would emerge. With only half of the design, the effects with the same patterns of signs are confounded.

The Terminology of Fractional Factorials

Effects with identical sign patterns are known as ALIASES. This is a very descriptive label, as it conveys the idea that a source may have more than one name. In our example A and BC are aliases, and so are the pair B and AC and the pair C and AB respectively. Sources that cannot be estimated at all because the design simply eliminates the comparisons necessary to evaluate them are known as DEFINING CONTRASTS. The primary defining contrast, the one used to divide the design in half (ABC in our example), is known as the SPLITTING CONTRAST. This name is given because it is the source used to break up the factorial design.

As the design is cut in half by the splitting contrast, one would expect the other half (that is, the other four treatment combinations) also to yield a table that shows the identical pattern of confounding. The table is not identical, for the signs are reversed; but the confounding is. This means that either of the sets of (four) treatment combinations may be used as a half-replicate of the 2^3 design, and either half-replicate may be used to obtain estimates of the three main effects.

Multiple Defining Contrasts

In our example, a 2^3 design was cut in half, thus reducing it to the size of a 2^2 design. If the design were larger, it might be sensible to cut it in half again with another splitting contrast. This repeated halving is the key to the tremendous savings in experimental effort afforded by the fractional factorial design. With designs having a large number of factors, the gain is dramatic. For example, a 2^8 design generates 256 treatment combinations; but a one-sixteenth replicate (4 halvings) requires only 16 treatments yet provides estimates for all of the main effects.

Each halving means another degree of confounding and additional aliases for the sources of primary interest. Fortunately, we need not construct a table of sign patterns in order to find the sets of aliases. There is a simple algorithm for determining them. The rule is to multiply the source name by the defining contrast and drop squared terms. Thus in our example where the defining contrast is ABC, the alias of A must be $A \cdot ABC = A^2BC = BC$, which is what the table told us. With multiple defining contrasts, the multiplying rule is applied repeatedly to determine all of the aliases.

Complications mount, however, because among any two or more splitting contrasts there is a generalized interaction that is also a defining contrast. The generalized interaction is found by multiplying the splitting contrasts and canceling factors with even exponents, while keeping those with odd exponents. As the generalized interactions are defining contrasts, they too yield aliases, and these aliases can be determined via the multiplying rule.

The generalized interaction makes the choice of splitting contrasts a nontrivial issue. It is all too easy to choose a set of contrasts that inadvertently loses a main effect by making it a generalized interaction. Typically, the researcher wants to adopt a plan in which sources of interest (usually main effects) are confounded with sources thought to be negligible (usually high-order interactions). Because determining such plans is a tedious, trial-and-error process for all but half-replicates, standard research practice is to look for a suitable plan in chapter 6A of Cochran and Cox (1957).

Computations

The way to analyze a fractional factorial is to pretend that the design is a complete factorial of lower order. Thus, half of a 2^P can be viewed as a 2^{P-1}, while a quarter of a 2^P can be analyzed as a 2^{P-2}. For example, suppose we have a one-fourth replicate of a 2^5 design, with the contrasts ABE and CDE used to split the design. The eight treatment combinations (taken from Cochran and Cox) are:

11111, 22111, 11221, 21212, 12212, 21122, 12122, 22221.

If we ignore factors D and E, the treatment combinations are 111, 221, 112, 212, 122, 211, 121, 222. These eight sets of indices form a complete 2^3 design, although they are not ordered properly. To do the analysis, we rearrange the scores so that the indices for A, B, and C are in the proper order for a three-factor design, then proceed with an ordinary ANOVA. Each of the seven degrees of freedom is identified with one of the sources from the five-factor design, as are the associated mean squares. The identification requires the choice of one of the four aliases in each case. These aliases are obtained from the three defining contrasts: ABE, CDE, and ABCD (the latter is the generalized interaction obtained by multiplying ABE by CDE). The list of sources should look like this:

2^3 ANOVA (Preliminary)	Aliases	Source (Final Table)
A	BE, ACDE, BCD	A
B	AE, BCDE, ACD	B
C	ABCE, DE, ABD	C
AB	E, ABCDE, CD	E
AC	BCE, ADE, BD	AC or BD
BC	ACE, BDE, AD	BC or AD
ABC	CE, ABDE, D	D

The column labeled "Source" gives the identification that the researcher would apply to each of the confounded sources. In general, main effects have first priority, followed by interactions in increasing order of complexity. As always, the researcher must justify the assumption that the other aliases are not contributing to the sum of squares.

Error Terms

The sources are then tested against whatever error term is in use for the design. Any of the customary approaches to generating an estimate of variability may be used with fractional factorial designs. For a within-cells error term, the design is administered to several sets of subjects. For a repeated-measures design, each of the subjects gets all or some part of the fractional replicate. In general, the fractional replicate is regarded as though it were a complete design.

For an example of a fractional factorial design, let us consider an experiment with seven factors, each one having two levels. A complete design would have 128 cells, but here a one-eighth replicate will be used, so only 16 treatment combinations will be presented.

In a study of attitudes toward smoking, panelists were asked to give their degree of agreement to the proposal that furnishing cigarettes to a minor should be treated as a felony (0 = extreme disagreement, 100 = extreme agreement). Panelists were classified according to gender (G), whether they were smokers (S), whether a parent had died of lung cancer (D), age (A—whether they were over or under 40), and to political party affiliation (P—Republican or Democratic). Each panelist was shown one of two films (F); the severe film showed rotting lungs and coffins, while the mild film featured graphs and panel discussions. Half of the panelists were also offered a free month's supply of cigarettes (C). Three panelists were obtained for each of the sixteen cells in the design, so there are 48 responses. The three contrasts used to split the design were GSDA, GDPC, and GSPF. Here are the data:

G	S	D	A	P	F	C	
Female	Nonsmoker	No Death	Under	Democratic	Mild	No cigarettes	78, 89, 56
Male	Smoker	Death	Over	Democratic	Mild	No cigarettes	49, 62, 68
Male	Smoker	No Death	Under	Republican	Severe	No cigarettes	42, 15, 31
Male	Nonsmoker	Death	Under	Republican	Mild	Cigarettes	18, 33, 26
Male	Nonsmoker	No Death	Over	Democratic	Severe	Cigarettes	55, 41, 52
Female	Smoker	Death	Under	Democratic	Severe	Cigarettes	12, 17, 31
Female	Smoker	No Death	Over	Republican	Mild	Cigarettes	19, 45, 36
Female	Nonsmoker	Death	Over	Republican	Severe	No cigarettes	8, 3, 19
Male	Smoker	No Death	Under	Democratic	Mild	Cigarettes	92, 67, 87
Male	Nonsmoker	Death	Under	Democratic	Severe	No cigarettes	75, 58, 52
Male	Nonsmoker	No Death	Over	Republican	Mild	No cigarettes	38, 29, 41
Female	Smoker	Death	Under	Republican	Mild	No cigarettes	21, 27, 18

G	S	D	A	P	F	C	
Female	Smoker	No Death	Over	Democratic	Severe	No cigarettes	56, 61, 47
Female	Nonsmoker	Death	Over	Democratic	Mild	Cigarettes	39, 34, 43
Female	Nonsmoker	No Death	Under	Republican	Severe	Cigarettes	25, 51, 44
Male	Smoker	Death	Over	Republican	Severe	Cigarettes	3, 18, 12

The first step in the analysis is to determine all of the defining contrasts. The number of contrasts indicates the extent of confounding, and the set of contrasts determines the aliases for each source. The additional defining contrasts are obtained by pairwise multiplication of the splitting contrasts; also, the product of all of the splitting contrasts is a defining contrast.

$$GSDA \times GDPC = SAPC$$

$$GSDA \times GSPF = DAPF$$

$$GDPC \times GSPF = DCSF$$

$$GSDA \times GDPC \times GSPF = GACF$$

For a one-eighth replicate, there are seven defining contrasts in all. They will be used to determine which terms will be identified as the seven main effects and eight interactions in the ANOVA table.

To generate the table, we must order the treatment combinations so that the responses are in the appropriate order for a 2^4 design with 3 scores per cell. First, the four factors to be used in the preliminary analysis must be chosen; any set of four factors that do not form a defining contrast may be used. Notice that the first four factors as given in the data set would not be a suitable choice, because GSDA is a defining contrast. The second through fifth factors, SDAP, are convenient to use, and so we write their indices for each of the treatment combinations. Of course, it does not matter which level of each factor is designated as level 1 or level 2, so long as we are consistent.

S	D	A	P	Scores	Factorial Order
2	2	2	2	78, 89, 56	16
1	1	1	2	49, 62, 68	9
1	2	2	1	42, 15, 31	7
2	1	2	1	18, 33, 26	6
2	2	1	2	55, 41, 52	12
1	1	2	2	12, 17, 31	13

S	D	A	P	Scores	Factorial Order
1	2	1	1	19, 45, 36	3
2	1	1	1	8, 3, 19	2
1	2	2	2	92, 67, 87	15
2	1	2	2	75, 58, 52	14
2	2	1	1	38, 29, 41	4
1	1	2	1	21, 27, 18	5
1	2	1	2	56, 61, 47	11
2	1	1	2	39, 34, 43	10
2	2	2	1	25, 51, 44	8
1	1	1	1	3, 18, 12	1

The column labeled "factorial order" refers to the standard arrangement for computer entry: 1111 is first, 2111 is second, 1211 is third, etc. The resulting ANOVA table is in terms of S, D, A, and P.

Source	df	SS	MS	F
S	1	105.02	105.02	<1
D	1	4237.52	4237.52	39.98*
A	1	728.52	728.52	6.87
P	1	10179.20	10179.20	96.03*
SD	1	99.19	99.19	<1
SA	1	999.19	999.19	9.43*
SP	1	13.02	13.02	<1
DA	1	336.02	336.02	3.17
DP	1	20.02	20.02	<1
AP	1	15.19	15.19	<1
SDA	1	713.02	713.02	6.73*
SDP	1	368.52	368.52	3.48
SAP	1	426.02	426.02	4.02
DAP	1	1716.02	1716.02	16.19*
SDAP	1	875.52	875.52	8.26*
Within cells	32	3392.00	106.00	

It only remains to identify the sources in the table with the most preferred of their seven aliases. As a general rule, one should assume that terms of lower order (i.e., main effects or two-factor interactions) are more likely to be contributing to the sum of squares than their higher-ordered counterparts, and so three-factor and four-factor interactions should be replaced if possible. In the absence of psychological theorizing, all two-factor interactions have an equal claim on their confounded sum of squares, so the other two-factor aliases for each should be given. In the final table, then, the sources may be given as (the aliases enclosed by parentheses would not be included—they are given to help you locate the correct term):

Source	df	SS	MS	F
S	1	105.02	105.02	<1
D	1	4237.52	4237.52	39.98*
A	1	728.52	728.52	6.87*
P	1	10179.20	10179.20	96.03*
G (= SDA)	1	713.02	713.02	6.73*
F (= DAP)	1	1716.02	1716.02	16.19*
C (= SAP)	1	426.02	426.02	4.02
SD = GA = CF	1	99.19	99.19	<1
SA = GD = PC	1	999.19	999.19	9.43*
SP = GF = AC	1	13.02	13.02	<1
DA = GS = PF	1	336.02	336.02	3.17
DP = GC = AF	1	20.02	20.02	<1
AP = SC = DF	1	15.19	15.19	<1
GP = DC = SF (= SDAP)	1	875.52	875.52	8.26*
SDP = ···	1	368.52	368.52	3.48
Within cells	32	3392.00	106.00	

The sources separated by equals signs have equivalent status in the table, while those in parentheses would be regarded as replaced. The three-factor term at the bottom of the table, SDP, has seven three-factor aliases, all of which have equivalent status. Since a heavily confounded, high-order interaction is not likely to be interpretable, I have not bothered to list the aliases.

Evaluation of Fractional Factorial Designs

Practical application of fractional designs has been limited for several reasons. Certainly, construction and analysis are more difficult than for complete designs, and the capability of ordinary computer programs for factorial ANOVA to handle these fractions is not widely known. Researchers with a superficial acquaintance worry about confounding sources of interest with interactions of unknown magnitude.

However, the alternative to a fractional design is usually not a complete factorial incorporating the factors of interest. It is a reduced complete factorial, one that simply omits factors of possible substantive value. The researcher constrained by pragmatic concerns is motivated to ignore variables whose contribution is uncertain, so that the project is of manageable size. Consider, for example, the classic experiments during the golden age of social psychology (roughly 1957–1967, before ethics codes imposed limits on the imagination). Many of the most prominent of these studies ignored subject gender and experimenter gender. Usual practice was either to use only one gender or simply not to keep track of these classifying variables that more recent research has shown to be so powerful. A carefully chosen fractional design can allow the examination of the main effects and low-order interactions of potentially interesting variables without making the experiment impossibly large.

The obvious danger, confounding important effects so they cannot be estimated purely, cannot be ignored. If effect A is confounded with effect BC, then the source identified as A may reflect any kind of combination of the two sources. A nonsignificant effect for A may be indicative of an ineffectual variable, or it may mean that BC has cancelled out the effect of a potent A. A general solution to this problem is available if the same variables A, B, and C are embedded in a new (or repeated) fractional factorial design. By reversing the assignment of stimuli to the levels of A, so that the previous A_1 becomes A_2 and vice versa, the researcher changes the direction of the effect. The confounding that previously produced cancellation of effect should now yield augmentation.

Exercises

13-1. A psychology major conducted a study of reading speed using his roommate as the subject. The dependent variable was the number of pages read during a one-hour test period. Three kinds of reading material were used (an introductory psychology text, a science fiction novel, a mystery). Three different drinks preceded the reading hour (coffee, milk, beer). The reading hour itself was either 9 a.m., 2 p.m., or 11 p.m. A complete factorial design could not be used because the roommate was willing to devote only nine hours to the study. Analyze the data.

Drink	Reading Material		
	Psychology text	Science fiction novel	Mystery novel
Coffee	9 a.m. 45	2 p.m. 80	11 p.m. 73
Milk	2 p.m. 37	11 p.m. 64	9 a.m. 81
Beer	11 p.m. 30	9 a.m. 52	2 p.m. 56

13-2. A comparative psychologist wished to study the effects of four different reinforcers on learning rates in four species of quadruped. The task required a discrimination response—one bar press or two, depending on which of the two stimuli was presented, and the dependent variable was the number of learning trials required until the animal got ten in a row correct. Having recently read about hemispheric differences in the brain, the psychologist incorporated an additional factor into the study, namely, which limb the animal must use for the response. Only eight animals of each species were available. Analyze the results and discuss the reasonableness of the experimenter's implicit assumption. Notice that each score is from a different animal, since carry-over effects would be expected here.

	Limb			
Reinforcer	Right forepaw	Left forepaw	Right hind leg	Left hind leg
Sugar pellet	goats 30, 42	cows 20, 17	horses 42, 23	pigs 11, 15
Ear rub	pigs 35, 38	horses 36, 42	cows 83, 64	goats 27, 38
Banana	cows 52, 83	pigs 14, 26	goats 56, 23	horses 44, 53
Pepsi	horses 32, 49	goats 51, 34	pigs 23, 35	cows 37, 48

13-3. Imagine that World War II has just broken out and it is vital to national security to increase production. At the factory to which you have been called, production is measured in terms of rivets riveted. Management has tried five ideas to boost riveting, and your job is to evaluate them.

The ideas proposed have been sensible: increasing pay (I), showing films of concentration camps (F), playing drum-like music during riveting hours (D), giving a free war bond to the worker riveting the most (W), and injecting workers with caffeine every morning (C). The industrial psychologist who planned the study would have liked to conduct a complete factorial design to test these five ideas, but the pressure of the situation did not allow enough time since the company had only eight riveting squads, each composed of five workers. In fact, before the psychologist could process the data, she was drafted. She left behind the data layout for the vice president in charge of testing to fill in, along with a mysterious note that said "the splitting contrasts are IFC and DWC." Help to win the war by analyzing the data (the number of rivets riveted during the test week).

Squad 1—no new ideas	Squad 2—I, F
342, 515, 247, 404, 312	649, 575, 890, 423, 509
Squad 3—D, W	Squad 4—I, D, C
448, 302, 215, 416, 502	942, 816, 743, 829, 716
Squad 5—F, D, C	Squad 6—I, W, C
780, 404, 319, 714, 638	800, 754, 619, 793, 911
Squad 7—F, W, C	Squad 8—I, F, D, W
814, 518, 932, 665, 782	602, 479, 318, 465, 503

13-4. The following data come from a length estimation experiment. Each of the eight subjects saw 16 sequences of six lengths. At each of the six serial positions, the stimulus presented was either a level 1 (80 mm) or a level 2 (120 mm). At the end of each sequence presentation, the subject estimated the average length in millimeters. The sequences were constructed to form a one-fourth replication of a 2^6 design. Analyze the data. The splitting contrasts were ABCE and ABDF.

Responses

				Subjects				
Sequences	PK	SP	SR	JG	RB	CS	PC	CC
111111	95	80	90	95	80	90	90	70
111212	105	90	105	120	75	115	110	75
112121	110	80	95	125	65	80	95	80
112222	105	115	90	135	125	120	100	90
121122	115	115	105	100	90	115	110	90
121221	95	95	110	95	90	95	110	100
122112	90	105	90	120	105	110	95	70
122211	90	105	95	100	80	75	115	80
211122	90	100	115	105	85	115	100	55
211221	110	80	105	145	85	115	100	85
212112	100	100	100	110	95	115	105	80
212211	100	95	115	90	105	120	105	60
221111	85	75	90	100	70	95	100	70
221212	110	115	110	130	125	115	120	100
222121	100	105	110	110	115	125	100	85
222222	115	130	110	135	125	145	120	100

13-5. Four drugs were contrasted in a study employing eight participants, each of whom received all of the drugs in a controlled sequence. Sequence was deemed relevant because interactions between successively administered drugs are always a consideration in such research. Four orderings were used, with two participants assigned to each. Evaluate the treatments effect and any other sources you judge worthy of attention. The scores are response times to a mock traffic light (in milliseconds). Roman numerals denote the sequential position of the treatment for each participant pair.

		Treatment			
Ordering		T_1	T_2	T_3	T_4
O_1		III	IV	I	II
	P_1	135	189	214	156
	P_2	158	211	198	177
O_2		II	III	IV	I
	P_3	129	173	235	164
	P_4	146	159	178	155
O_3		I	II	III	IV
	P_5	98	175	190	162
	P_6	125	198	175	138
O_4		IV	I	II	III
	P_7	133	164	158	167
	P_8	122	195	216	150

(Hint: This problem is confusing because ordering and sequence are distinct variables. Ordering is a grouping factor, with participants nested under it. This nesting will lead you to the error term for ordering. Sequence, like treatments, is a variable of substantive interest. Keep in mind that the substantive effects, including the residual, are based on the 16 cell totals [and so comprise 15 df], and the error terms will account for the remaining 16 df.)

Answers to Exercises

13-1.

Source	df	SS	MS	F
Drink	2	643.56	321.78	3.80
Reading material	2	1872.89	936.45	11.05
Time	2	20.22	10.11	<1
L.S. Error	2	169.55	84.78	

13-2.

Source	df	SS	MS	F
Limb	3	1289.84	429.95	3.30*
Reinforcer	3	2064.84	688.28	5.28*
Species	3	2716.84	905.62	6.94*
Residual	6	1462.45	243.74	1.87
Within cells	16	2087.50	130.47	

The implicit assumption is that limb, reinforcer, and species do not interact. I know too little about animal behavior to do more than raise the question.

13-3.

Source	df	SS	MS	F
Increasing pay	1	235162.20	235162.20	12.59*
Films	1	3115.22	3115.22	<1
Drum-like music	1	42445.22	42445.22	2.27
War bond	1	126.02	126.02	<1
Caffeine	1	721728.30	721728.30	38.64*
FD = IW	1	78057.23	78057.23	4.18*
ID = FW	1	2002.23	2002.23	<1
Within cells	32	597676.00	18677.38	

13-4.	Source	df	SS	MS	F
	Subjects	7	11073.20	1581.89	
	A	1	1098.63	1098.63	5.97*
	Error (AS)	7	1287.30	183.90	
	B	1	775.20	775.20	4.02
	Error (BS)	7	1348.24	192.61	
	C	1	468.95	468.95	3.27
	Error (CS)	7	1004.49	143.50	
	D	1	2583.01	2583.01	81.56*
	Error (DS)	7	221.68	31.67	
	E (= ABC)	1	1547.07	1547.07	29.76*
	Error (ABCS)	7	363.87	51.98	
	F (= ABD)	1	3559.57	3559.57	16.95*
	Error (ABDS)	7	1470.12	210.02	
	AB = CE = DF	1	297.07	297.07	3.29
	Error (ABS)	7	632.62	90.37	
	AC = BE	1	328.32	328.32	1.60
	Error (ACS)	7	1438.87	205.55	
	BC = AE	1	1.76	1.76	<1
	Error (BCS)	7	427.93	61.13	
	AD = BF	1	508.01	508.01	10.08*
	Error (ADS)	7	352.93	50.42	
	BD = AF	1	4.88	4.88	<1
	Error (BDS)	7	756.06	108.01	
	CD = EF	1	142.38	142.38	1.20
	Error (CDS)	7	831.06	118.72	
	DE = CF (= ABCD)	1	0.20	0.20	<1
	Error (ABCDS)	7	823.24	117.61	
	ACD = BDE = BCF = AEF	1	431.45	431.45	5.51
	Error (ACDS)	7	548.24	78.32	
	BCD = ADE = ACF = BEF	1	43.95	43.95	<1
	Error (BCDS)	7	1910.74	272.96	

13-5.	Source	df	SS	MS	F
	Orderings (groups)	3	2124.84	708.28	2.20
	Error (P + PO)	4	1288.88	322.22	
	Treatments	3	19619.59	6539.86	16.11*
	Sequences	3	344.34	114.78	<1
	Residual	6	1304.73	217.46	<1
	Error	12	4870.63	405.89	

Note: Some authorities recommend pooling residual and error to get increased df for error, but usually it is preferable to maintain the residual to furnish a partial check on the key assumption of negligible interaction.

14

Introduction to Functional Measurement

Functional measurement is foremost an approach to understanding cognitive behavior. Because the methodology was developed using ANOVA as the primary analytic technique, a brief explanation is given here. One goal of this presentation is to inspire the reader to view analysis of variance as more than a way to confirm differences among groups. The two volumes by Anderson (1981, 1982) will provide those whose interest has been stimulated with the information needed to carry out a full-scale functional measurement study.

Cognitive algebra is the heart of the method. A theoretical model is proposed for a behavioral task. The task always involves joint consideration of two or more stimuli or stimulus aspects. Thus, a subjective integration is called for. Usually the model is phrased as a simple equation. For example, a person may be asked to envision the attractiveness of a meal consisting of a sandwich and a drink (Shanteau & Anderson, 1969). As a working hypothesis, one might expect the appropriate model to be a weighted combination of the attractiveness of the individual components. The components will certainly have differing values across individuals. For me, tuna in pita gets a positive evaluation and cranberry juice a negative one; but your taste may be quite different. In the same way, weights for the components may differ. For me, the sandwich is important and the drink is almost an afterthought. By presenting factorial combinations of various hypothetical sandwiches and drinks, the researcher elicits ratings that may be analyzed for conformity with the hypothesized model. The ratings may then be decomposed to yield estimates of the subjective attractiveness of the individual sandwiches and drinks. These estimates differ from those obtained with simple ratings in that their validity is established via the prior verification of the model.

The capability of these methods to transcend face validity is a great advance for psychological theory. Functional measurement has been applied to research domains ranging from psychophysics (Anderson, 1970) to marital interaction (Anderson & Armstrong, 1989).

The creative step in a functional measurement analysis is the selection of an appropriate judgmental task. This includes finding a way to elicit quantitative responses to the stimulus combinations that will be presented. The task is tied to an algebraic model whose correctness is an empirical matter. If the data support the proposed model, scaling may proceed. This is the sense in which the measurement is functional.

Of course, an algebraic model has an "as-if" character. It is not claimed that a person truly punches buttons on an internal calculator. Rather the mental processes are equivalent to mathematical operations, in the same way that aiming a football emulates calculations of arcs and forces. We need not impute use, or even knowledge, of calculus to a quarterback who accurately controls the ball's trajectory.

The Functional Measurement Diagram

The fundamental problem for psychological measurement, as opposed to physical measurement, is that the quantities we wish to measure are unobservable. We must infer rather than directly assess. The intensity of a tone can be measured with an instrument. The loudness of that sound, however, its subjective intensity, is concealed in the observer's head. Similarly, when a person integrates several subjective intensities, the result of that combining is internal. Any evaluation we might extract from the person is external, and we must be concerned with possible distortion introduced during the translation from internal representation to overt response.

With an established model in hand, functional measurement has the leverage to solve simultaneously the three key equations underlying a set of judgments. These equations can be seen in the FUNCTIONAL MEASUREMENT DIAGRAM (Anderson, 1981):

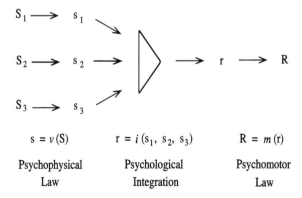

Functional Measurement Diagram

In the DIAGRAM, capital letters refer to observables, small letters to unobservables, and italicized letters to functions. The researcher's primary goal is to estimate the functions. Here, there are three stimuli that give rise to one response. The main interest is usually in the integration function, *i*, but all of the functions may have substantive interest.

The PSYCHOPHYSICAL LAW connects a subjective value (s_i) to each objective stimulus (S_i). While the search for this relationship has been considered to be within the domain of psychophysicists since the time of Fechner (Krueger, 1989), its ramifications extend to all behavioral research. How can we know if the subject extracts the same meaning from a stimulus that the researcher intends it to contain? Not to assess this connection in a research context is to assume a linear relation between stimulus and sensation.

The bulk of the research on the psychophysical law has been carried out via direct measurement, in which the subject's expression of sensation is taken at face value. In contrast, functional measurement derives the psychological law from the integration rule, whose validity can be assessed objectively.

PSYCHOLOGICAL INTEGRATION refers to the combining rule used by the subject to put together the internal representations of the stimuli. For example, we might propose that the subjective area of a rectangle is the product of subjective width and subjective height. That is hardly a startling hypothesis, although unexpectedly it is not quite correct empirically (Anderson & Weiss, 1971). Less obviously, perhaps, a similar multiplying model has also been used to describe the fear felt by a person who has a phobia of snakes as the product of the snake's valence (picture, rubber snake, or real snake) and the likelihood of exposure.

For most of the judgmental tasks that have been studied, simple algebraic models such as adding, averaging, or multiplying have been found to provide fairly good descriptions of the integration rule. Such rules are amenable to verification via ANOVA, and that has been a factor in the choice of task. Surprisingly, however, complex tasks that seem as though they ought to call for complex processes often do not. Sometimes this occurs because a complex strategy turns out to be mathematically equivalent to a simple one. In other cases, we flatter ourselves that our judgments are intricate and deep, when in reality they can be described by simple equations. A subject's verbal statement of his integration rule is only a datum, and it is no more than a possible starting point in functional measurement. Certainly there are judgmental tasks whose complexity exceeds the descriptive power of a simple algebraic rule, but it has been a delightful surprise to see the broad range over which these models have been successful.

The PSYCHOMOTOR LAW is the output function by which the result of the integration gets translated into an overt response. If the output function is nonlinear, then the raw responses will not exhibit the predicted pattern even though the proposed model is true at the level of the internal response. Nonlinear transformation alters *F* ratios, and so a correct model would appear incorrect because the crucial ANOVA test would fail.

On the other hand, successful verification of the proposed model immediately confers validity on the overt response scale. If the response scale were inducing distortions, then the model would appear to fail. Therefore, the overt responses must be linearly related to internal judgments. This is perhaps the major theoretical contribution of functional measurement, that it provides a means of validating responses for what may superficially look like a standard questionnaire-type instrument. For a typical questionnaire used by a social researcher, each response is an independent entity. Consequently, validity is largely in the eye of the beholder.

In contrast, functional measurement considers the set of responses. Success of the model is seen from the pattern within the set, and this success in turn validates the response scale. The logic seems a little circular (and is not quite impeccable), but it is good enough for practical scientific inference. The possibility does exist that an incorrect model and an invalid response scale might just happen to nullify each other's impact. In practice, researchers do not allow for the occurrence of this remarkable coincidence.

Functional measurement allows for transformation of the response, in an attempt to apply m^{-1} (the inverse of the m function) and thereby undo the effect of the psychomotor law. If it is possible to find a transformation that brings about the desired conformity, then the model is deemed correct and the transformed response scale is held to be the valid one. In some cases, one may anticipate that transformation will be required. In bisection, for example, the response is on the stimulus continuum. Nonlinear output thereby expresses the effect of Fechner's law. Transformation is needed to demonstrate the expected additive integration rule (Weiss, 1975).

The majority of functional measurement studies employ numerical responses, which have the virtue of simplicity and are familiar to most subjects. It is not difficult for a well-read analyst to find evidence that such instruments may be subject to bias. What prevents a researcher from transforming ad lib and thereby confirming an incorrect model? There are two checks upon the freedom to transform. First, the successful transformation must be MONOTONE; that is, it must preserve the order of the original responses. If the internal response, r, is not ordinally related to R, the overt response, then the response instrument is too chaotic to salvage. Second, there should be convergence across judgmental tasks that use the same response instrument (Birnbaum & Veit, 1974). The responses from the other tasks should require the same transformation to establish their integration rules.

In general, though, it is preferable to bypass the need for transformation by setting up conditions that will produce a linear psychomotor law. Careful experimental technique is a cornerstone (Anderson, 1982). Technique includes such mundane matters as allowing sufficient practice and making sure the participants understand the instructions. It also extends to theoretically interesting issues such as establishing end-anchors for the response scale or avoiding response instruments known from previous research to be biased. End-anchors should be established while the instructions are given or during practice trials. By presenting stimuli for which the expected responses are more extreme than any presented in the actual design, the researcher tries to induce the subject to use the entire response scale in a consistent manner. Numerical responses that are not numerical from the subject's perspective (such as marking a line that will subsequently be measured by the researcher) eliminate number biases and minimize the ability to recall previous answers.

General Strategy

First, we examine the general strategy for a functional measurement study, and then we look at differences in the analyses for various models. We shall consider three (the most prominent three) types of integration rule: adding, multiplying, and averaging. The discussion will be confined to judgments of two stimuli (two-factor designs). While there is no theoretical limit on the number of factors in a design, the principles can be seen for these relatively simple cases. The FUNCTIONAL MEASUREMENT computer program allows up to four stimulus factors,[1] along with subject and/or replicate factors.

Cognitive algebra is concerned with the integration of information. In the integration process, each subjective value (s_i) in a stimulus array is weighted by its salience or importance (w_i) and the products are then combined to yield an overall judgment (r). The overt counterparts of a set of such judgments (the Rs) are decomposed by the researcher via the verified model to allow estimation of the subjective values.

Factorial Design

The set of stimulus arrays to be judged is constructed according to a factorial design. The factors are chosen for their substantive interest; the researcher thinks they will combine in an interesting way. The particular levels of the factors, the actual stimuli to be presented, are seldom of great import. The levels should be chosen to span their respective factors as fully as is practical. If there is interest in mapping a psychophysical law for the dimension, then the set of levels will have numerical values. For other applications, the levels may be ordinal or even nominal. The combinations resulting from crossing the levels constitute the stimulus arrays to be judged. These combinations are presented in a random order, perhaps with distracter items to keep the subject from responding according to a pattern rather than expressing true judgments.

Ecological validity is inevitably a concern with stimuli constructed artificially. The researcher must find combinations of factors that make sense to the subject. The task of rating a meal consisting of a sandwich plus a drink seems sensible to me. Consider, though, the structurally equivalent task of judging the quality of an experience that includes reading a book while having a drink. The experimenter could vary the books and vary the drinks, but the integration by the subject could proceed, in my opinion, only in a formal way because the judgment is not a natural one. The data might well be consistent with an additive model, but the cognitive process is likely to be a laboratory creation that sheds little light on everyday thinking.

One way to avoid artificial tasks is to seek out stimuli with a natural, inherent factorial structure. Ideally, the stimulus combinations will appear as unitary, albeit multifaceted, entities. For example, Rundall and Weiss (1994) asked nurses to rate how afraid of contagion they would be if they were caring for patients with various diseases. The diseases varied in transmissibility and in severity of prognosis, factors one would expect to contribute to the perceived danger. The tricky element for the researchers was to find actual diseases conforming to a complete factorial design.

The algebraic models make specific predictions about the analysis of variance of the factorial design. The predictions have the form that certain sources will exist and others will not. In practice, a prediction of nonexistence is considered to be confirmed when the corresponding F ratio proves nonsignificant. These predictions also have visual counterparts in terms of a pattern that should appear in the factorial plot.

Typically, main effects will have huge F ratios; these merely indicate that the factors are making their expected contributions. The significance of main effects must be confirmed but is not of great interest. A nonsignificant main effect can mean that the factor does not play a role in the judgment, so that the model is incorrect. More likely, though, the levels are simply not subjectively different, reflecting a flaw in experimental technique. Normally, such difficulties would be exposed during pilot work.

The crucial tests, those that require careful examination, center on interaction. The existence of an interaction (or in some cases, of a specified component of an interaction) jeopardizes the model. From a statistical perspective, there is a danger here, in that one is trying to show that a predicted source yields a nonsignificant F ratio. To be meaningful, the goodness-of-fit test requires the analysis to have sufficient power to detect a true violation. Large main effect F ratios are encouraging, but are not sufficient to guarantee power to find interactions. Insisting that the data appear to be in general visual accord with the predicted pattern helps to guard against Type II errors. The problem is diminished in designs with several factors, because the emergence of an occasional significant interaction sustains confidence in the statistical power of the analysis (Weiss & Anderson, 1969). I note in passing that researchers using correlation-based techniques for evaluating models have incorrectly looked at proportions of variance. If main effects account for a large proportion, then automatically interaction accounts for a small proportion and is likely to be dismissed, thereby prompting acceptance of the model. However, the magnitude of main effects depends heavily on the spacing of the levels and is unrelated to the issue of deviations from the proposed model (Anderson & Shanteau, 1977).

A model is usually held to apply at the level of the single subject. Accordingly, analysis is carried out for individual subjects when there is enough data to yield reasonable statistical power. The single-subject design calls for replication within each subject, so that a within-replicates mean square may be used as the error term. In some cases, researchers have carried out single-subject analyses in which the stimuli have been tailored (Jaccard & Becker, 1985) to the individual subject. These personal designs (Anderson, 1990) are especially suitable when individuals can be expected to have dramatically different views of particular stimuli; for example, males and females are likely to have different emotional responses to the prospect of breast cancer. Within-subject replication is probably best avoided when the subject can recall specific responses to specific stimulus combinations; in such cases, the customary independence assumption is violated. The Show Panels for third factor option is convenient for examining single-subject data; let subjects be that third factor.

When individual analysis is impossible (as is the case when there is only one replicate per subject), the model must be evaluated at the level of the group. Group analysis can also provide a convenient summary after individual tests have been carried out. Certainly, the familiar caveat that a summary curve based on averages does not necessarily reflect the average of the individual curves is in order. On the other hand, if the model is correct for individuals it is likely to be correct for the group as well. An advantage of the group analysis is that systematic violations are likely to be detected, even if they are too small to show up in individual analyses. Of course, estimation of group subjective values may not be of great interest unless uniformity across individuals has been shown.

The group design entails each subject going through the entire factorial arrangement one or more times. With subjects as a factor in the design, group analyses employ standard repeated-measures error terms; each source is tested against its interaction with subjects. Use of repeated-measures error terms for the group analysis has the logical consequence that model evaluation for the group can yield a different outcome from that expected from evaluations of individual-subject data.

A problem arises if it is deemed impractical to run each participant through the entire factorial design. Nesting subjects under some of the treatment combinations is a possibility. A potential drawback, though, is that the likely mixture of idiosyncratic scale values and cognitive strategies across individuals would in some cases render the model analysis meaningless.

The nested design may be more apt when the grouping variable is a natural one not subject to experimental control, such as gender or ethnicity. Rundall and Weiss (1998) asked outpatients with chronic diseases to project their compliance with hypothetical therapeutic medications that produced various side effects. Patients were nested under two specific characteristics associated with their diseases, symptom severity and prognosis. For example, hypothyroidism is a disease with moderate symptoms and a favorable prognosis, while hypertension is asymptomatic but has an unfavorable prognosis. The model sought to describe the way disease features and side effects combine to determine compliance. Because the individual's experience of disease features is limited to a specific disease, the model can be defined only at the overall group level.

Another possibility is the use of a true independent groups design, one in which each subject produces only one response. The researcher might be concerned that exposure to one stimulus combination could prevent the subject from considering any other combination independently. Although the model and scale values are evaluated only at the level of the group, the use of random assignment allows the locus of the model to be an abstract "average" subject. For discussion and illustration of the independent groups design, see Howe (1991).

Scaling

A hidden assumption built into the goodness-of-fit test is that of independence of scale values. The subjective value of cranberry juice does not depend on whether it is paired with tuna or with hamburger. Should the scale value instead fluctuate with context, that would be an interaction and accordingly the model would be incorrect.

After the model passes the goodness-of-fit test, estimation of the subjective values is straightforward. Values are defined on either an individual or a group basis. The marginal means provide optimal estimates on an interval scale for each factor separately. For each level, then, the corresponding marginal mean is the subjective value. These means may be reported as functional scale values.

Researchers often elect to incorporate a rescaling operation. Recall that an interval scale is unique up to a linear transformation and thus the origin and unit are arbitrary (e.g., Centigrade is a linear transformation of Fahrenheit). An alternative way of looking at an interval scale is that if any two values are assigned, then all of the others are determined. The rescaling option offered by FUNCTIONAL MEASUREMENT utilizes this characteristic of the interval scale in allowing the analyst to fix any two values. Rescaling is not a necessary step in finding the subjective values. The adjustment merely affords the convenience of placing the subjective values on a numerical scale comparable to that of the objective values. Rescaling is most likely to be of interest when the stimuli are specified numerically.

Comparison of scale values across the two factors is only possible with special designs. If determination of the psychophysical law for a particular factor is important, then a goodly number of levels should be used.

Transformation

It should be stressed that the normal, or default, analysis simply uses the raw data without adjustment. Transformation may affect the outcome of the model evaluation. Nonlinear transformation will alter scale value estimates and the error distribution as well as F ratios. On the other hand, because linear transformation does not affect F ratios, it may be applied freely for convenience; one might wish to have the responses confined to a particular range. For the purpose of model evaluation, though, a linear transformation is merely an unnecessary effort.

Specific nonlinear transformations may be employed if the researcher has reason to believe the response scale is distorted in a particular way. For example, the arcsin may be called for with percentage scores known to be compressed at high values. A priori transformation with a theoretical basis is always allowable.

In contrast, post-hoc nonlinear transformation is dangerous and should call for a df penalty (perhaps a reduction of the df_{error}) to compensate for possible capitalization on chance effects. The magnitude of that penalty has not been studied. It is usual in the goodness-of-fit trade to subtract 1 df for each estimated parameter, and so it would be natural to suggest a 1 df penalty for a function such as the logarithm. However, if a researcher cycles through various transformations looking for the one that achieves the intended variance reduction, then a larger penalty seems in order. This concern applies with even more force to the more vigorous transformations designed to achieve maximal reduction of specified sources, such as FUNPOT (Weiss, 1973) or MONANOVA (Kruskal, 1965). Transformation may be planned in advance, employing the theoretical justification that the researcher expects the response scale to induce distortion of the additive internal responses. However, because the exact form of the transformation depends upon the data, the procedure must be regarded as post-hoc. FUNPOT incorporates a *df* penalty, subtracting 1 *df* from the source(s) reduced for each degree of the sufficient polynomial. MONANOVA utilizes an unknown number of *df*.

Specific Models

Adding

Certainly the simplest and probably the most commonly used model in psychological research, the adding model states that a combined judgment is formed by summation. The internal response is the sum of the weighted subjective values of the components. The formal equation is:

$$r = w_A s_{A_i} + w_B s_{B_j} + e_{ij} \qquad (14\text{-}1)$$

Here, e_{ij} is the random component with which we are familiar; e_{ij} has an expected value of zero. We can explore the predictions of this model by looking at the expected values of the internal responses for a 2 (rows) × 3 (columns) design:

		Column stimuli			
		s_{B_1}	s_{B_2}	s_{B_3}	Marginal mean
Row stimuli	s_{A_1}	$w_A s_{A_1} + w_B s_{B_1}$	$w_A s_{A_1} + w_B s_{B_2}$	$w_A s_{A_1} + w_B s_{B_3}$	$w_A s_{A_1} + (w_B/3)(s_{B_1} + s_{B_2} + s_{B_3})$
	s_{A_2}	$w_A s_{A_2} + w_B s_{B_1}$	$w_A s_{A_2} + w_B s_{B_2}$	$w_A s_{A_2} + w_B s_{B_3}$	$w_A s_{A_2} + (w_B/3)(s_{B_1} + s_{B_2} + s_{B_3})$

The expected response in each cell may be seen to be the sum of the weighted subjective counterparts, s_{A_i} and s_{B_j}, of the row stimulus (S_{A_i}) and the column stimulus (S_{B_j}). The weight for all of the row stimuli is the same value, w_A, and equally the weight for the column stimuli, w_B, is fixed. The weights are nonnegative but there is no constraint linking a pair.

It is easy to derive the graphical prediction of parallelism and the statistical prediction of no interaction; the difference in the predictions between for each pair of row entries is the same $(w_A s_{A_1} - w_B s_{A_2})$ across columns. Note that the parallelism does not require knowledge of the scale values but results only from the adding operation.

The marginal means for the row factor may be rewritten as $k_1 s_{A_1} + k_2$ and

$k_1 s_{A_2} + k_2$, where k_1 and k_2 are constants according to the model. This rewriting is the basis of the claim that the marginal means are proportional to the subjective values. Similar calculations for the marginal means associated with the columns have been left for the reader to carry out.

A subtracting model is formally identical to an adding model, and all of the predictions have the same form. One need only change the sign of the composition operation in each cell. The structural equivalence is easy to see if we envision that changing the sign of the operation is algebraically equivalent to changing the sign of the scale values on the subtracted factor. A subtracting model might be appropriate for a task such as allocating praise to persons who have had varying degrees of success (Praiseworthiness = Accomplishments − Luck).

Let us work through a numerical example. The example is artificially small, in that more subjects, more replicates, and perhaps a larger stimulus design would be used in a realistic experiment.

Our judgmental task is the evaluation of how intimidating a guard dog would be. The researcher hypothesizes that this quality depends primarily upon two attributes of dogs, the size and the loudness of the bark. Further, it is anticipated that these qualities will add, because either a large outline or a loud sound might serve to frighten a potential intruder. The attributes would seem to contribute independently to the dog's fearsomeness.

Of course, this model may not tell the whole story. Other characteristics, such as aggressiveness or readiness to bark might be important. In the experiment, the researcher would attempt to hold constant other relevant factors; for example, one might choose to have the dogs be of similar breed or appearance. This kind of control is extremely important, but the researcher may not have sufficient empirical knowledge to get the details right. Pilot work is usually needed. I bequeath to any readers who are dog experts the right to improve upon and actually carry out the investigation.

Raters might be chosen from among people who have gone shopping for guard dogs. Each would be presented a series of photographs showing a pooch next to a baby (to lend realism as well as to furnish a size reference) accompanied by recorded barks. Three sizes and three intensities would be crossed to yield nine dogs to be judged twice by each subject. The term *intensity* is used, rather than *loudness*, to describe the stimulus because intensity is a physical unit, and that is what the experimenter controls. The rater experiences loudness, which is intensity's subjective counterpart. For most continua, everyday language does not afford separate terms for the objective and subjective units, so the distinction brought out in the functional measurement diagram should be maintained by using an adjective.

The response scale is a 100-mm line labeled "terrible guard dog" at the left end and "great guard dog" at the right end; the individual judgments can be measured to yield scores between zero and 100. Preliminary training would include exposure to dogs expected to be rated better and worse respectively than any in the factorial design.

220   ANALYSIS OF VARIANCE AND FUNCTIONAL MEASUREMENT

The data from three raters might look like this:

	Rater 1			Rater 2			Rater 3		
Intensity	Small	Medium	Large	Small	Medium	Large	Small	Medium	Large
1	16	37	52	8	25	49	17	26	32
	22	31	45	15	33	52	10	23	45
2	48	52	78	52	64	73	31	39	53
	40	68	71	50	52	71	25	50	59
3	72	85	96	66	79	88	49	57	69
	65	75	91	72	86	93	54	67	68

The first step in the analysis is to evaluate the fit of the model for each subject separately. This is done by looking at an abbreviated version of the three ANOVAs.

Rater 1			Rater 2			Rater 3		
Source	df	F	Source	df	F	Source	df	F
Size	2	32.64*	Size	2	51.97*	Size	2	26.99*
Intensity	2	89.46*	Intensity	2	180.32*	Intensity	2	61.16*
SI	4	0.13	SI	4	2.80	SI	4	.64
Within	9		Within	9		Within	9	

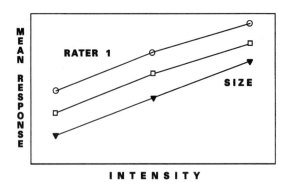

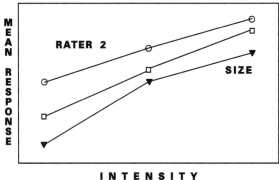

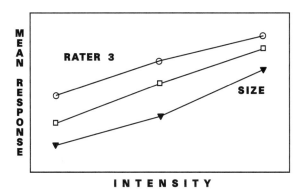

The parallelism in the figures, along with the lack of significant interaction, implies (subject to reservation based on the low power inherent in a small data set) that an additive model is appropriate for these data. The fit justifies the use of the marginal means as functional scales for size and for loudness of bark as given below.

For most of the continua we scale, standard language does not provide distinct terms for the stimulus domain and its subjective counterpart. The vocabulary of psychoacoustics is richer, though, and we follow convention in referring to the stimulus value as intensity and the sensation value as loudness.

Rater 1		Rater 2		Rater 3	
Size	Loudness	Size	Loudness	Size	Loudness
43.8	33.8	43.8	30.3	31.0	25.5
58.0	59.5	56.5	60.3	43.7	42.8
72.2	80.7	71.0	80.7	54.3	60.7

As the model seems to fit all of the subjects, the group analysis runs little risk of distortion and provides a compact summary of the data. The group curves appear parallel, and the group ANOVA is also consistent with an additive model.

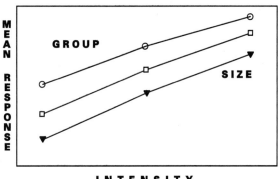

Source	df	SS	F
Raters	2	2549.5	
Size	2	6214.7	256.77*
RS (error)	4	48.4	
Intensity	2	17574.4	81.73*
RI (error)	4	430.1	
SI	4	161.2	1.85
RSI (error)	8	174.0	

Group	
Size	Loudness
39.6	29.9
52.7	54.2
65.8	74.0

Averaging

An averaging model has the same form as an adding model, except that the sum of the weights for a judgment is equal to one. This constraint can change the character of the integration. Consider the case in which S_A is evaluated very positively. If S_B is valued positively but less so than S_A, in an averaging model the judgment of the pair will be less than the value of S_A alone. In an adding model, though, the addition of any positively valued stimulus must raise the judgment.

There are two kinds of averaging model to consider. The simpler is the constant-weight averaging model. The weight for each of the factors is fixed no matter what level of the stimulus is presented. This model generates parallelism and nonsignificant interaction just as the adding model does. Both adding and constant-weight averaging are additive processes, so scaling is straightforward after the model is verified. However, the two kinds of additive integration cannot be distinguished with the usual designs. Instead, the level "none" must be used on one of the factors to assess the impact of mildly positive and mildly negative information.

In contrast, the differential-weight averaging model is more complex and much harder to evaluate. Here, the weight used for each stimulus varies with its subjective value. Parallelism is not expected. This case can be quite difficult to analyze and we shall not deal with it here.[2] An illustration of the complexity can be seen in Surber's (1984) detailed analysis of how people infer ability given information about performance and effort, and how they infer effort given information about performance and ability. For either judgment, the implied reliability of the information provided was varied, in a successful attempt to manipulate the weight the information would be given.

Multiplying

Multiplying models arise frequently in cognitive algebra. The usual way to envision them is to see that one variable modulates or magnifies the effect of another. Since weights ordinarily have a multiplicative function, they can be omitted from the model specification. We can conceptually regard the product of weight and scale value for each variable as the subjective value, s. The formal equation for the internal response is:

$$r = s_{A_i} s_{B_j} + e_{ij} \qquad (14\text{-}2)$$

We can derive the expected values of the internal responses to a factorial design as we did before:

		Column stimuli			
		s_{B_1}	s_{B_2}	s_{B_3}	Marginal mean
	s_{A_1}	$s_{A_1} s_{B_1}$	$s_{A_1} s_{B_2}$	$s_{A_1} s_{B_3}$	$s_{A_1}(s_{B_1} + s_{B_2} + s_{B_3})/3$
Row stimuli	s_{A_2}	$s_{A_2} s_{B_1}$	$s_{A_2} s_{B_2}$	$s_{A_2} s_{B_3}$	$s_{A_2}(s_{B_1} + s_{B_2} + s_{B_3})/3$
	s_{A_3}	$s_{A_3} s_{B_1}$	$s_{A_3} s_{B_2}$	$s_{A_3} s_{B_3}$	$s_{A_3}(s_{B_1} + s_{B_2} + s_{B_3})/3$

As was the case with the additive model, the marginal means for the multiplying model are linearly related to the internal values of the stimuli if the model is correct. Here each marginal mean may be written as $s_{A_i} k$.

Another striking property can be seen in the algebra. The responses across each row are the product of a constant (s_{A_i}) times a variable (s_{B_j}) that changes with the column. Thus they each have the form of the equation for a straight line passing through the origin. The lines differ only in their slopes—the row constants. If we were to graph the responses, we should have a set of straight lines intersecting at the origin and diverging as they move outward. This pattern is known as a linear fan.

In practice, though, two tricky elements need to be kept in mind. The first is that the quantities in the equation are unobservable. This means that simply graphing the set of responses against the stimulus values for one of the factors will not in general produce the appropriate pattern, because we don't know the abscissa values to use in the graph. Fortunately the marginal means come to the rescue, for we can see that they are linearly related to the internal values corresponding to the stimuli. Therefore if we plot responses using marginal means spacing (a functional scale, if you will) then the fan should appear. This argument holds even if the overt stimuli are only nominally scaled, or if the researcher cannot specify their proper ordering in advance. The subjective values to which the stimuli give rise are the quantities that enter the cognitive computations, and these subjective values are numerical by definition.

The second element is also easy to deal with. The origin for the response scale, the true zero, is not a defined quantity on an interval scale. This means that we cannot look for the lines to intersect at the origin unless the response scale has a meaningful zero. It is the pattern, the set of lines diverging from a common point, which announces the presence of a multiplying model.

The linear fan has a statistical counterpart that thereby allows a quantitative evaluation of the extent to which nonmultiplicativity is present in the data. The multiplying model implies that there will be interaction between the two factors. More specifically, the interaction will be concentrated in a single *df* component, the bilinear component. This bilinear, or linear × linear, component is a two-dimensional analog of the linear trend component of a main effect (cf. chapter 8). Subjective spacing, utilizing the marginal means, determines the orthogonal trend coefficients.

The significance testing procedure is analogous to that for a trend test. We must show that the hypothesized component is significant and the remainder is not. Here the focus is on the interaction. The bilinear component of the interaction represents the linear fan pattern and the residual the deviations from linearity. In order for a line to be defined, a factor must have at least three levels. The more levels on each factor, the greater the power to detect violations.

Our example of the multiplying model is taken from a study conducted by Anderson and Shanteau (1970) on the value of risky outcomes. Subjects were asked to evaluate the worth of gambles that varied in the probability of winning and the amount to be won. For example, how much is the gamble < 2/6 to win $.25 > worth to you? Responses were made on an unmarked rating scale anchored by −$.50 and $.50 from the subject's perspective; the range of possible responses as read by the researcher was −500 to 500. The bets were varied in a 5 (probabilities) × 4 (winnings) design, with levels 1/6, 2/6, 3/6, 4/6, 5/6 and $.05, $.10, $.25, and $.50. Subjects were run through four replications. We shall examine the data from two of the subjects.

Subject 1

Probability	Winnings	Replicate			
		1	2	3	4
1/6	$.05	42	40	50	64
1/6	$.10	12	26	72	22
1/6	$.25	100	128	20	42
1/6	$.50	60	168	6	104
2/6	$.05	12	34	44	48
2/6	$.10	60	46	88	110
2/6	$.25	78	98	118	98
2/6	$.50	114	180	146	74
3/6	$.05	116	102	70	38
3/6	$.10	108	90	86	122
3/6	$.25	174	226	258	234
3/6	$.50	242	264	354	270
4/6	$.05	66	112	60	76
4/6	$.10	118	168	110	104
4/6	$.25	222	240	226	194
4/6	$.50	310	352	334	396
5/6	$.05	104	86	64	54
5/6	$.10	108	110	138	110
5/6	$.25	222	250	234	254
5/6	$.50	408	456	480	500

Subject 2

Probability	Winnings	Replicate			
		1	2	3	4
1/6	$.05	2	−34	−16	−46
1/6	$.10	−10	−18	−14	−40
1/6	$.25	−12	−26	−4	6
1/6	$.50	100	−50	−6	40
2/6	$.05	30	−6	36	−48
2/6	$.10	32	−16	−24	−2
2/6	$.25	120	96	50	96
2/6	$.50	52	204	94	46
3/6	$.05	16	−22	56	74
3/6	$.10	30	68	60	30
3/6	$.25	132	160	112	168
3/6	$.50	242	160	170	222
4/6	$.05	80	64	28	92
4/6	$.10	60	60	44	36
4/6	$.25	120	140	184	168
4/6	$.50	356	372	362	346
5/6	$.05	52	38	44	54
5/6	$.10	80	94	48	94
5/6	$.25	182	184	218	222
5/6	$.50	360	450	466	432

The hypothesized model was simply a subjective version of the expected value of the gamble; the response should be the product of the subjective probability and the subjective value. This is an example of the use of a normative model (what a "correct" subject should do) as a starting point for understanding the behavior. It would not be surprising if subjective and objective values did not coincide, so we should not be surprised if the ratings do not match expected values. However, if an integration rule other than multiplication were found to apply, that would be a phenomenon worthy of explanation.

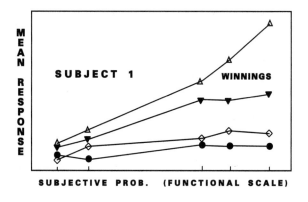

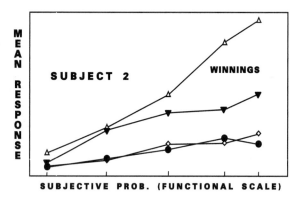

The plots show the data for the individual subjects. They seem generally to show the linear fan pattern, albeit with some irregularities for the lower payoffs, especially for subject 2. The spacing of the ticks on the horizontal axis conveys the subjective values used for determining coefficients. If the model is correct, these values are scale values corresponding to the objective values presented. The ANOVAs produce the same picture of the model's success. The multiplying model fits subject 1; but subject 2's residual, while small, is significant.

Subject 1			Subject 2		
Source	df	F	Source	df	F
Winnings	3	141.95*	Winnings	3	148.48*
Probs	4	72.74*	Probs	4	89.76*
(WP	12	14.52)	(WP	12	15.29)
L × L	1	155.74*	L × L	1	160.56*
Res	11	1.68	Res	11	2.08*
Within	60		Within	60	

Since the model fits subject 1, we can proceed to examine functional scales for that subject. Because there are numerical objective values against which to compare the subjective values, the obtained scales have intrinsic interest. To make the comparison easier, I used the rescaling option in the FUNCTIONAL MEASUREMENT program, linearly transforming the marginal means while forcing the first two values to match the objective values.

Subject 1

Winnings	Marginal mean	Rescaled	Probabilities	Marginal mean	Rescaled
$.05	64.10	5	1/6	59.75	1
$.10	90.40	10	2/6	84.25	2
$.25	170.80	25.29	3/6	172.13	5.59
$.50	260.90	42.41	4/6	193.00	6.44
			5/6	223.63	7.69

It may be seen that S_1 underestimates the value of $.50 and overestimates the higher probabilities relative to the values of 1/6 and 2/6 (accurate values would be 3.00, 4.00, and 5.00).

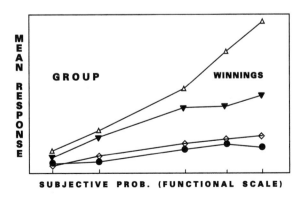

This minigroup may be examined for illustrative purposes (obviously two subjects are not really enough to constitute a group). The group figure on p. 227 shows a linear fan except for the problem that the two lower winnings values did not yield divergent curves. The ANOVA suggests that a multiplicative model is an adequate group description. Note that the kind of variability used as a yardstick here, inconsistencies between subjects, differs from the within-cell variability used in a single-subject analysis. So it is possible that a model will pass one kind of test and fail another.

Group

Source	df	SS	F
Subjects	1	94284.1	
Winnings	3	975784.3	300.25*
SW (error)	3	3249.9	
Probabilities	4	721533.4	75.24*
SP (error)	4	9589.4	
(WP	12	386515.2	25.13)
(S × WP	12	15383.8)	
L × L	1	351125.1	142848.50*
S × L × L	1	2.5	
Res	11	35390.1	2.30
S × Res	11	15381.1	

Now that the model is supported on a group basis, we can look at the group scale values. These differ from those for subject 1 alone. The group overvalues $.25 and $.50 relative to $.05 and $.10 (perhaps because the latter amounts are both seen as very small), and they overestimate the higher probabilities less than subject 1 did.

Group

Winnings	Marginal mean	Rescaled	Probabilities	Marginal mean	Rescaled
$.05	44.40	5	1/6	25.88	1
$.10	60.50	10	2/6	65.88	2
$.25	143.30	53.74	3/6	138.50	4.31
$.50	240.90	106.84	4/6	175.00	5.21
			5/6	206.13	6.12

In this study, carried out when functional measurement was young, the normative model was apparent to both the researcher and the subject. And though the stimuli were unitary entities, the formal, mathematical nature of the task was even more pronounced in the examination of apparent size of rectangles (Anderson & Weiss, 1971). The obvious nature of the correct approach to the task and the transparency of the factorial design typified those early studies, as researchers were learning to work with algebraic models. When Rundall and Weiss (1994) explored nurses' fear of contagion, the judgmental task was more holistic. Nurses were asked how fearful they would be of providing care for a patient with particular disease characteristics. The multiplicative model was conceived by analogy, viewing the provision of care as a gamble. Disease transmissibility—how likely is the nurse to catch the disease—is akin to probability, and the severity of the disease determines how much the nurse "wins" (or in this case, loses).

The Challenge of Choosing and Implementing a Model

Where does a model come from? One week's mail suggested a cognitive operation that might be interesting to analyze. It contained appeals from several worthy charitable organizations. Common to them was a description or photo of something in terrible jeopardy along with an assurance that my money could help to alleviate the problem. Endangered were (a) the Bill of Rights, (b) starving children, (c) elephants, (d) whales, (e) baby fur seals, and (f) the homeless. What determines the likelihood that a potential donor will contribute?

It is clear that the charities think an emotional message is paramount. The pictures of the seals being clubbed would break anyone's heart. The organizations also are sympathetic to my financial straits. They tell me that the requested contribution is really only "pennies a day" or the "cost of a candy bar," and that it is tax deductible. What they don't express very clearly, though, is just how my forgoing that candy bar will prevent those pathetic seals from becoming coats.

Other elements certainly enter the decision to contribute. The donor's current financial commitments and political philosophy surely affect responding. These aspects are relatively immune to manipulation by a small brochure. Let us focus, then, on two elements that may be the most susceptible to a well-phrased appeal; the amount of emotionality inspired and the level of confidence that the donation will advance the donor's cause. How do these combine? Inquiring charitable organizations want to know.

My guess is that emotional arousal and faith in the contribution's effectiveness combine multiplicatively. The basis of the guess is that if either were zero, no response would be forthcoming. Even if tremendous guilt is aroused and I am moved to tears, there is no point in just throwing money away. I must believe that some alleviation of the problem will occur if I contribute. In an additive process, a high value on either variable brings about a large response; the variables operate independently. For a multiplying process, a low value on either variable produces a low yield. To generate a large outcome, both variables must have high values.

The distinction between models has an important implication for the way in which charities go about their fund-raising. Standard procedure seems to presume that pathos is the key; give the donor the maximum tolerable dose. This strategy presumes an additive model (or perhaps more accurately, ignores the other factor in what I am suggesting is a multiplicative process). For some organizations, the practical limit on emotionality may have already been reached. If that is so, additional efforts aimed at previously canvassed targets will not prove very rewarding.

On the other hand, if my unsubstantiated proposal is correct, a road to more effective campaigns is indicated. The appeals need to be rewritten to emphasize efficacy. I must be shown how my particular contribution will accomplish my goal. The religious group that sends donors a picture of the specific children their donations have "saved" seems implicitly aware of the multiplying nature of the charitable decision.

How can the proposed model be evaluated? Setting up the requisite factorial design would appear to be straightforward. A 5×5 design for saving the whales would involve manipulating expressions of the sadness of their plight and the likelihood of saving them via donations.

One problem, though, is that it might not be possible to manipulate the two variables independently. It may be that the perceived severity of the situation affects how the efficacy information is accepted ("if the whales are in such bad shape, maybe nothing can be done to save them despite the apparent efficiency of this foundation").

Another, more general, problem concerns implementation in a realistic way. Should the elicited response be monetary or should it be a rather more vague rating of willingness to contribute? The model is presumed to function at the individual level. Yet it would hardly be practical to ask participants to even look at the 25 stimulus pairings, let alone to try to extract monetary pledges from them. One might attempt to bypass the difficulty by defining the cognitive unit of interest as the group, thereby presenting one pairing to each subject. This would allow a realistic experiment, but variation across individuals in subjective values and in financial status would enter the error terms and contribute to low power.

Solving these kinds of problems is a principal joy of research. Functional measurement, like any vibrant paradigm, calls for imaginative implementation. There is plenty of room left for progress at both the theoretical and applied levels.[3] Think, as well as read, and have fun.

Use of the FUNCTIONAL MEASUREMENT Program

FUNCTIONAL MEASUREMENT carries out most of the analyses needed for additive or multiplicative models on a single-subject and/or group basis. Two sample data files using the data from examples in this chapter, have been provided on your CD. They are ADDSAMP.DAT, the guard dog data (arranged as 3 sizes $\times$ 3 intensities $\times$ 2 replications $\times$ 3 judges), and MULTSAMP.DAT, the gambling data (arranged as 4 winnings × 5 probabilities × 4 replications × 2 subjects).

The program begins with data input, for which the scheme is consistent with that used by FACTORIAL ANOVA. The user transmits information about the levels for the substantive factors and the number of replicates (how many times each subject goes through the design).

The transformation option should be exercised during model confirmation only if theory specifies the expected form of nonlinear distortion in the response scale.[4] On the other hand, transformation may be freely used for exploratory purposes with an eye toward future model confirmation using an independent data set. For an additive model, post-hoc transformation using either MONANOVA or FUNPOT is available. The goal of both of these algorithms is to make the data as additive as possible; MONANOVA constrains the transformation to be monotone, while FUNPOT does not. In either case, the additive data obtained must be regarded with caution, as the statistical consequences of transformation have not been explored. If one's view is that the program's transformation has merely undone a transformation imposed by the response instrument, then a tentative proposal that the underlying cognitive process is additive may be in order. Here is the design specification for the guard dog data, presuming we want to carry out a FUNPOT transformation (in reality, no transformation is needed because the raw data are already additive):

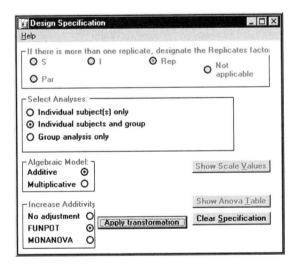

The default options for FUNPOT are usually adequate. In this example, there is no need to specify the source to be reduced; because there are only two substantive factors, it is their interaction that will be reduced.

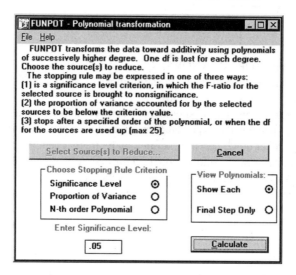

If you are testing a multiplying model, you will be asked how you want to construct the orthogonal coefficients needed for the bilinear analysis. Coefficients should normally be based on the marginal means, since the means are the best estimates of the subjective values entering into the hypothesized internal multiplication. You can look at the coefficients if you like, though I confess I've never found much meaning to them. Group coefficients are used for the group ANOVAs (whether the design is independent groups or repeated-measures), while the individual's coefficients are used for single-S analyses. Here is the design specification for the gambling data:

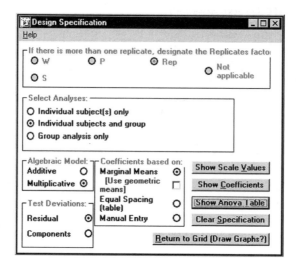

ANOVAs are the next offering. If you are carrying out the bilinear analysis, you can select the components to be inspected. The usual model evaluation requires that the linear × linear component be significant and the remainder of the interaction, or the residual, be nonsignificant. Whether the model tested is additive or multiplicative, if it passes the test then scale values are meaningful. The program offers this option independently of whether the model fits, as you may wish to examine the scale values for exploratory purposes. For quantitative factors, it may be useful to rescale.

To carry out linear rescaling of the marginal means, you choose the values to be assigned to any two levels on each factor; usually the values chosen are the objective values. This option would be useful if you had a theoretical reason for fixing the values; for example, the responses to monetary stimuli can be set to look like a utility scale with money-valued anchors. It is customary to set either the first two levels or the first and last. Adjustments made by rescaling are only cosmetic (because subjective scales are interval scales, which are not fundamentally altered by linear transformations) and do not affect model tests at all.

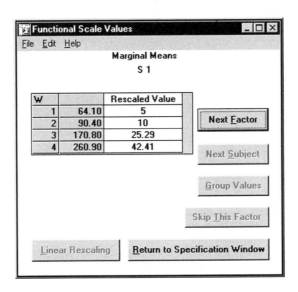

Factorial plots showing how the two factors combine may be constructed with FUNCTIONAL MEASUREMENT. You may choose which factor you wish to see on the horizontal axis. The other factor will appear as the parameter in the graph. The points on each curve may be spaced according to the marginal means (the usual choice for a multiplying model), or spaced equally (normal for additive models), or spaced arbitrarily (for whatever creative reason you dream up); this choice is accessible through the X-Axis Options Menu. You may get as many plots as you wish for a data set. Plots for individual subjects may be obtained using the Show Panels for third factor button.

Exercises

14-1. Mother always said it was just as easy to fall in love with a rich woman. "Otherwise, my son, you'll spend your days sitting at a computer typing meaningless numbers." At the Alumni Dating Service, clients are shown a photo and told the occupation of prospective mates. They are then asked to express on a ten-point rating scale their degree of interest in the person. Two photos are shown per weekly session, so over a four-month period 32 prospective mates can be shown to a client. This happy arrangement led to the construction of a 4 (photos) × 4 (occupations) set of potential mates, presented twice to each client in a random order except for the restriction that the same photo could not be shown both times during a session.

Does an additive model describe the evaluations? Can you develop a scale for occupations (L = lawyer, A = accountant, P = psychologist, B = bureaucrat)? Here are the responses from three clients, all of whom were shown the same stimulus set.

	Client 1 Photo				Client 2 Photo				Client 3 Photo			
	1	2	3	4	1	2	3	4	1	2	3	4
L	4	7	5	8	1	1	2	3	7	5	5	10
	5	6	3	8	2	1	1	4	8	3	5	10
A	2	3	4	6	6	5	3	8	6	4	5	9
	4	4	2	7	5	7	5	9	5	7	3	8
P	7	8	6	9	4	4	3	7	8	7	6	8
	5	8	7	10	4	6	2	6	8	6	4	9
B	3	4	4	5	7	6	4	9	5	2	4	6
	3	3	3	3	5	4	5	10	4	4	2	8

14-2. Students go to the Campus Employment Service, where there are jobs offering different pay rates and demanding different qualifications. What determines an applicant's choice? Although most students think money is a good thing, they do not all apply for the highest-paying job. One reason for this reluctance is that the chances of being an acceptable candidate are not equal across jobs. What seems to occur is a kind of matching, whereby the students go for the highest-paying job for which they feel qualified.

To study this matching process, a researcher used a computer to generate stimulus pairs that were judged by student subjects who were asked to imagine they were applying for a job. The computer program asked for the applicant's qualifications (experience, skills, courses) and then constructed job requirements that matched those individual qualifications to varying degrees. For example, a student who spoke Spanish and knew accounting would find some jobs had precisely those requirements while other jobs called for very different abilities. One factor in the design, then, was the extent to which qualifications met requirements; the levels were low, medium, and high. The other factor was the starting pay. The hourly levels for pay were $4, $6, $8, and $10. The two factors were crossed to yield 12 hypothetical jobs, and these were presented twice. The students were asked to indicate how likely they would be to apply for each job as it was presented by the computer. Responses were on an unmarked 100-mm line whose left end was marked "not at all likely" and whose right end was marked "certainly would apply."

The researcher expected a multiplying model would describe the way in which the two factors combined. Here are the data for three of the subjects. Is her expectation supported? Does the psychophysical scale for money seem reasonable?

Pay	Match								
	Subject 1			Subject 2			Subject 3		
	Low	Medium	High	Low	Medium	High	Low	Medium	High
$4	10	32	42	7	26	56	13	25	38
	6	29	49	10	29	49	9	32	45
$6	14	39	57	5	33	68	11	47	63
	8	35	61	10	45	72	14	53	67
$8	17	58	77	11	51	84	14	65	81
	10	64	82	8	58	79	7	59	84
$10	15	73	98	14	69	96	8	79	97
	9	65	96	10	59	93	5	74	99

14-3. In an effort to gain understanding of the medical issue of quality of life, a researcher asked hospital patients with potentially fatal, painful conditions (cancer, heart disease, diabetes) to consider the trade-off between survival duration and pain. The patient was asked to make a judgment of the value of a hypothetical therapeutic procedure that would afford a specified degree of pain reduction but would also affect survival chances. The ratings were made in terms of the proportion of accumulated net worth (excluding personal residence) the patient would be willing to give up to achieve the specified outcome. All of the participants in the study were between 40 and 50 years old; this range was chosen so that survival durations of 30 years would be meaningful. A typical question was "What proportion of your net worth would you be willing to pay for a procedure that *completely eliminated* pain but resulted in an expected survival duration of *ten years?*" A 4 × 5 factorial design was used to construct the hypothetical outcomes.

Twenty-four patients were tested, but (to spare labor) you are asked to analyze the data from a randomly selected group of three. Because the study required patients to reflect on issues of life and death, it was decided to collect only one replication per individual. Determine an appropriate descriptive model for the integration of survival duration and pain. Develop subjective scales for the two attributes.

Patient 1

Expected survival duration	Pain reduction			
	None	Slight	Moderate	Complete
1 year	1	8	15	22
5 years	7	15	32	45
10 years	7	19	40	57
20 years	9	24	53	68
30 years	10	29	55	80

Patient 2

Expected survival duration	Pain reduction			
	None	Slight	Moderate	Complete
1 year	2	10	13	17
5 years	5	13	25	37
10 years	8	17	29	42
20 years	7	25	38	57
30 years	11	28	42	61

Patient 3

Expected survival duration	Pain reduction			
	None	Slight	Moderate	Complete
1 year	1	5	18	20
5 years	1	8	30	49
10 years	2	15	45	65
20 years	3	18	57	79
30 years	3	22	63	88

14-4. The normative model for the equisection task presented in exercise 6-4 is for the respondent to average the subjective darknesses of the two gray chips presented on each trial. Equal-weight averaging is an additive process, so the internal responses should be additive. But because equisection calls for the overt responses to be made on the physical continuum, the data will appear additive only if the relation between the physical continuum and its internal representation is linear. Psychophysical research has demonstrated that nonlinear relations are the norm for many of the customary stimulus domains—brightness and loudness are the prototypes. These relations are often referred to as psychophysical or Fechnerian functions, after the German scientist who first explored the issue.

If an additive model really does underly the equisection judgments, then it should be possible to undo the nonlinear transformation introduced via the response system by transforming the data to additivity. In doing so, we will have solved Fechner's classic problem. The transformation that undoes the nonlinearity must be the inverse of the function that describes how a stimulus is represented as a sensation. Use the FUNPOT algorithm to determine the post-hoc transformation that makes the exercise 6-4 data additive. If that transformation is monotone, it is a viable candidate for the psychophysical function.

14-5. Charlie the Tuna knows that good taste is important. But nowadays we also worry about vitamins, calories, and cholesterol. Should the Campus Cafeteria tell all? Propose a model and design an experiment that incorporates the impact of nutritional information along with flavor on the desirability of various foods.

14-6. Authorities from the Secretary of Education to Magic Johnson tell kids to stay in school. But we've also heard that experience is the best teacher. Do salaries depend upon educational attainment and experience? Design an experiment in which the participants simulate a personnel analyst whose task is to determine appropriate salaries for clerical workers of varying backgrounds. Propose a model for the judgments.

14-7. I'm sure you've spent many pleasurable hours studying statistics. As we come to the last exercise, it is fitting to reflect on the factors that determine the amount of time a student devotes to a particular class. Two obvious candidates are the difficulty of the material and its relevance to the student's interests. Relevance is not only personal, but also may depend upon whether the course is required, recommended, or purely elective. Propose a model and design an experiment to examine the decisions about how to allocate study time. Choose a subject population for which this question has been important.

Answers to Exercises

14-1. Before scaling, the model must be evaluated. We look at the ANOVAs.

Client 1			Client 2			Client 3		
Source	df	F	Source	df	F	Source	df	F
Photo	3	16.24*	Photo	3	24.03*	Photo	3	24.09*
Occupation	3	30.24*	Occupation	3	35.62*	Occupation	3	8.65*
PO	9	1.38	PO	9	1.29	PO	9	1.19
Within	9		Within	9		Within	9	

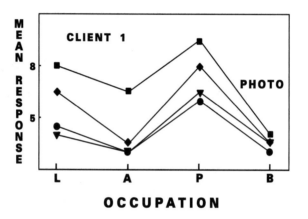

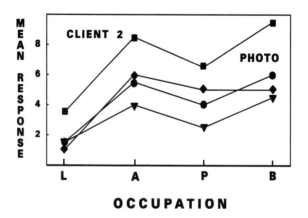

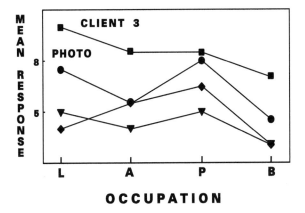

Although the ANOVAs for the individual clients satisfy the criteria for an additive model, I would be reluctant to pronounce these data additive. The individual factorial plots show worrisome crossover interactions, especially for clients 2 and 3. Clarification may come from the group analysis.

Group

Source	df	SS	F
Clients	2	27.9	
Photo	3	174.5	16.22*
CP (error)	6	21.5	
Occupation	3	41.2	<1
CO (error)	6	167.4	
PO	9	8.9	<1
CPO (error)	18	25.6	

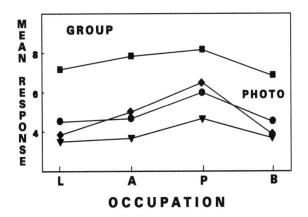

240 ANALYSIS OF VARIANCE AND FUNCTIONAL MEASUREMENT

The group data round out this picture of low power. The nonsignificance of the group job main effect (caused by differences in how the individual clients value the different occupations) and the crossovers in the group plot are sufficient to cause suspension of judgment. It would seem that more data per subject are needed and that the analysis will have to be presented on an individual basis. A possible locus of the difficulty is the 10-point response scale. The responses do not seem to have high within-cell variability; rather the problem may be that the scale is too constricted to allow effective expression of opinion. In collecting additional data, it might be worthwhile to explore an alternative response mode.

Scaling the occupations would be premature. I might add that this is a delicate decision. Generally, it is a good idea to be conservative in accepting a model. If you elected to accept for client 1 and reject for the other two, I would have no quarrel.

14-2. Model evaluation comes first. We look at the ANOVAs.

	Subject 1			Subject 2			Subject 3	
Source	df	F	Source	df	F	Source	df	F
Match	2	484.91*	Match	2	446.78*	Match	2	630.71*
Pay	3	82.74*	Pay	3	43.88*	Pay	3	95.26*
(MP	6	16.75*)	(MP	6	8.52*)	(MP	6	31.10*)
L × L	1	93.97*	L × L	1	43.55*	L × L	1	180.29*
Res	5	1.30	Res	5	1.52	Res	5	1.26
Within	12		Within	12		Within	12	

Everyone's ANOVA results satisfy the criteria for a multiplicative model, and the plots are consistent with the linear fan. Therefore it is appropriate to scale the pay variable. Use of the rescaling option allows the scale values to be easily compared to the objective pay offerings.

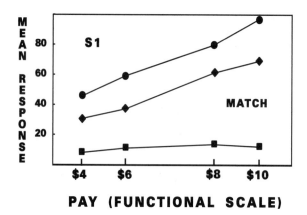

INTRODUCTION TO FUNCTIONAL MEASUREMENT

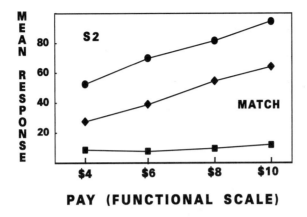

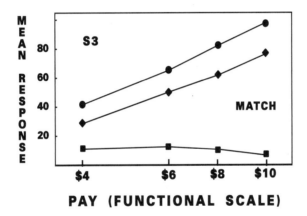

	Subject 1		Subject 2		Subject 3	
Pay	Marginal mean	Rescaled	Marginal mean	Rescaled	Marginal mean	Rescaled
$4	28.0	4	29.5	4	27.0	4
$6	35.7	6	38.8	6	42.5	6
$8	51.3	10.09	48.5	8.07	51.7	7.18
$10	59.3	12.17	56.8	9.86	60.3	8.30

The model fits the individuals, and so it is worth exploring whether a group analysis can provide a convenient summary description.

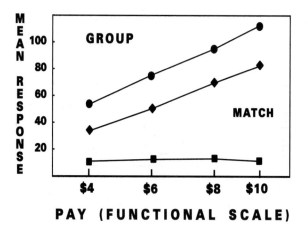

Group			
Source	df	SS	F
Subjects	2	56.6	
Match	2	47353.0	321.13*
SM (error)	4	294.9	
Pay	3	9682.4	108.74*
SP (error)	6	178.0	
(MP	6	4595.0	35.02)
(S × MP	12	262.4)	
L × L	1	4359.2	54.57*
S × L × L	2	159.8	
Res	5	235.8	4.60*
S × Res	10	102.6	

The failure of the model on a group basis precludes using group scale values. The conclusion would be that all three of the subjects multiply, but their subjective values don't agree closely enough for a group description to be appropriate.

14-3. This experiment can be analyzed only on a group basis, as there is but one replicate per subject. The appropriate model would seem to be multiplicative; the underlying intuition is that if no pain reduction were promised, what value would a long lifespan have?

Group

Source	df	SS	MS	F
Patients	2	382.03		
Duration	4	6767.67	1691.92	107.31*
PD (error)	8	126.13	15.77	
Reduction	3	19830.18	6610.06	38.40*
PR (error)	6	1032.77	172.13	
(DR	12	2757.40	229.78	22.56)
(P × DR	24	244.40	10.18)	
L × L	1	2715.13	2715.13	28.91*
P × L × L	2	187.84	93.92	
Res	11	42.27	3.84	1.49
P × Res	22	56.56	2.57	

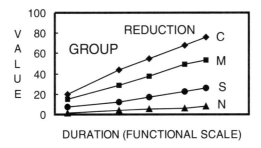

The model passes the test, and so scaling is justified. Since duration is measured on a quantitative scale, rescaling may make the values easier to understand. As one might expect, long survival durations are under-appreciated relative to the shorter ones. As pain reduction was presented in verbal terms, there was no reason to rescale the marginal means.

Duration	Marginal mean	Rescaled	Reduction	Marginal mean
1	11.00	1	None	5.13
5	22.25	5	Slight	17.07
10	28.83	7.34	Moderate	37.00
20	36.50	10.07	Complete	52.47
30	41.00	11.67		

14-4. A polynomial of degree 4 successfully reduced the Left by Right interaction. The transformed ANOVA passes the model test. The transformation appears to be a reasonably viable candidate for a proper psychophysical function throughout most of the range of the data, although there is a distinct departure from monotonicity for the lightest grays.

Source	df	SS	MS	F
Section	2	882.22	441.11	43.47*
Left	2	21170.20	10585.10	1045.54*
Right	2	9978.78	4989.39	492.82*
SL	4	537.03	134.26	13.26*
SR	4	42.23	10.56	1.04
LR	1	7.76	7.76	<1
SLR	8	471.48	58.93	5.82*
Within cells	81	820.05	10.12	

$$m^{-1} = R + (-.0392664)R^2 + .0007183R^3 + (-.0000047)R^4$$

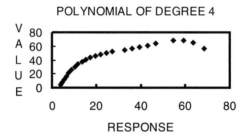

14-5, 14-6, and 14-7. I wish I could share your answers, but that pleasure will have to be reserved for another time. Discuss your proposals with colleagues. Please send me a copy of the article(s) when your research is published.

Notes

1. For an additive model, a standard ANOVA program such as FACTORIAL ANOVA can be used for the bulk of the quantitative analysis needed. FUNCTIONAL MEASUREMENT offers some extra options for additive models and is necessary for complete analysis of a multiplying model. The algorithms for the multiplying model were first developed for Group-Individual POLYLIN (Weiss & Shanteau, 1982).

2. Estimation of the large set of parameters required to fit the differential-weight averaging model is not a simple matter. A computer program, AVERAGE (Zalinski & Anderson, 1986), has been published but it is so unwieldy that I cannot recommend it.

3. Almost since the inception of functional measurement, there has been a competitor that also examines algebraic structure using factorial designs. Conjoint measurement, developed by Luce and Tukey (1964), is an axiomatic approach that has relied upon choice data rather than numerical responses. Although conjoint measurement has an elegant theoretical structure and was viewed favorably in an influential text (Krantz, Luce, Suppes, & Tversky, 1971), the approach has had little empirical success compared to functional measurement (Anderson, 2001).

4. An a priori transformation was employed by Weiss and Gardner (1979) in a study of subjective hypotenuse estimation. The Pythagorean theorem served as a normative model for the judgments. Because the square roots of the sums of the squares of the sides were supposed to be generated by an additive process, the raw responses were squared prior to analysis.

Appendix A: *F* Table

F TABLE

f_1 Degrees of Freedom (for greater mean square)

f_2	1	2	3	4	5	6	7	8	9	10	11	12	14	16	20	24	30	40	50	75	100	200	500	∞	f_2
1	161 4,052	200 4,999	216 5,403	225 5,625	230 5,764	234 5,859	237 5,928	239 5,981	241 6,022	242 6,056	243 6,082	244 6,106	245 6,142	246 6,169	248 6,208	249 6,234	250 6,261	251 6,286	252 6,302	253 6,323	253 6,334	254 6,352	254 6,361	254 6,366	1
2	18.51 98.49	19.00 99.00	19.16 99.17	19.25 99.25	19.30 99.30	19.33 99.33	19.36 99.36	19.37 99.37	19.38 99.39	19.39 99.40	19.40 99.41	19.41 99.42	19.42 99.43	19.43 99.44	19.44 99.45	19.45 99.46	19.46 99.47	19.47 99.48	19.47 99.48	19.48 99.49	19.49 99.49	19.49 99.49	19.50 99.50	19.50 99.50	2
3	10.13 34.12	9.55 30.82	9.28 29.46	9.12 28.71	9.01 28.24	8.94 27.91	8.88 27.67	8.84 27.49	8.81 27.34	8.78 27.23	8.76 27.13	8.74 27.05	8.71 26.92	8.69 26.83	8.66 26.69	8.64 26.60	8.62 26.50	8.60 26.41	8.58 26.35	8.57 26.27	8.56 26.23	8.54 26.18	8.54 26.14	8.53 26.12	3
4	7.71 21.20	6.94 18.00	6.59 16.69	6.39 15.98	6.26 15.52	6.16 15.21	6.09 14.98	6.04 14.80	6.00 14.66	5.96 14.54	5.93 14.45	5.91 14.37	5.87 14.24	5.84 14.15	5.80 14.02	5.77 13.93	5.74 13.83	5.71 13.74	5.70 13.69	5.68 13.61	5.66 13.57	5.65 13.52	5.64 13.48	5.63 13.46	4
5	6.61 16.26	5.79 13.27	5.41 12.06	5.19 11.39	5.05 10.97	4.95 10.67	4.88 10.45	4.82 10.29	4.78 10.15	4.74 10.05	4.70 9.96	4.68 9.89	4.64 9.77	4.60 9.68	4.56 9.55	4.53 9.47	4.50 9.38	4.46 9.29	4.44 9.24	4.42 9.17	4.40 9.13	4.38 9.07	4.37 9.04	4.36 9.02	5
6	5.99 13.74	5.14 10.92	4.76 9.78	4.53 9.15	4.39 8.75	4.28 8.47	4.21 8.26	4.15 8.10	4.10 7.98	4.06 7.87	4.03 7.79	4.00 7.72	3.96 7.60	3.92 7.52	3.87 7.39	3.84 7.31	3.81 7.23	3.77 7.14	3.75 7.09	3.72 7.02	3.71 6.99	3.69 6.94	3.68 6.90	3.67 6.88	6
7	5.59 12.25	4.74 9.55	4.35 8.45	4.12 7.85	3.97 7.46	3.87 7.19	3.79 7.00	3.73 6.84	3.68 6.71	3.63 6.62	3.60 6.54	3.57 6.47	3.52 6.35	3.49 6.27	3.44 6.15	3.41 6.07	3.38 5.98	3.34 5.90	3.32 5.85	3.29 5.78	3.28 5.75	3.25 5.70	3.24 5.67	3.23 5.65	7
8	5.32 11.26	4.46 8.65	4.07 7.59	3.84 7.01	3.69 6.63	3.58 6.37	3.50 6.19	3.44 6.03	3.39 5.91	3.34 5.82	3.31 5.74	3.28 5.67	3.23 5.56	3.20 5.48	3.15 5.36	3.12 5.28	3.08 5.20	3.05 5.11	3.03 5.06	3.00 5.00	2.98 4.96	2.96 4.91	2.94 4.88	2.93 4.86	8
9	5.12 10.56	4.26 8.02	3.86 6.99	3.63 6.42	3.48 6.06	3.37 5.80	3.29 5.62	3.23 5.47	3.18 5.35	3.13 5.26	3.10 5.18	3.07 5.11	3.02 5.00	2.98 4.92	2.93 4.80	2.90 4.73	2.86 4.64	2.82 4.56	2.80 4.51	2.77 4.45	2.76 4.41	2.73 4.36	2.72 4.33	2.71 4.31	9
10	4.96 10.04	4.10 7.56	3.71 6.55	3.48 5.99	3.33 5.64	3.22 5.39	3.14 5.21	3.07 5.06	3.02 4.95	2.97 4.85	2.94 4.78	2.91 4.71	2.86 4.60	2.82 4.52	2.77 4.41	2.74 4.33	2.70 4.25	2.67 4.17	2.64 4.12	2.61 4.05	2.59 4.01	2.56 3.96	2.55 3.93	2.54 3.91	10
11	4.84 9.65	3.98 7.20	3.59 6.22	3.36 5.67	3.20 5.32	3.09 5.07	3.01 4.88	2.95 4.74	2.90 4.63	2.86 4.54	2.82 4.46	2.79 4.40	2.74 4.29	2.70 4.21	2.65 4.10	2.61 4.02	2.57 3.94	2.53 3.86	2.50 3.80	2.47 3.74	2.45 3.70	2.42 3.66	2.41 3.62	2.40 3.60	11
12	4.75 9.33	3.88 6.93	3.49 5.95	3.26 5.41	3.11 5.06	3.00 4.82	2.92 4.65	2.85 4.50	2.80 4.39	2.76 4.30	2.72 4.22	2.69 4.16	2.64 4.05	2.60 3.98	2.54 3.86	2.50 3.78	2.46 3.70	2.42 3.61	2.40 3.56	2.36 3.49	2.35 3.46	2.32 3.41	2.31 3.38	2.30 3.36	12
13	4.67 9.07	3.80 6.70	3.41 5.74	3.18 5.20	3.02 4.86	2.92 4.62	2.84 4.44	2.77 4.30	2.72 4.19	2.67 4.10	2.63 4.02	2.60 3.96	2.55 3.85	2.51 3.78	2.46 3.67	2.42 3.59	2.38 3.51	2.34 3.42	2.32 3.37	2.28 3.30	2.26 3.27	2.24 3.21	2.22 3.18	2.21 3.16	13

F TABLE—(Continued)

f_2									f_1 Degrees of Freedom (for greater mean square)													f_2
	1	2	3	4	5	6	7	8	9	10	11	12	14	16	20	24	30	40	50	75	100	200	500	∞	
14	4.60 8.86	3.74 6.51	3.34 5.56	3.11 5.03	2.96 4.69	2.85 4.46	2.77 4.28	2.70 4.14	2.65 4.03	2.60 3.94	2.56 3.86	2.53 3.80	2.48 3.70	2.44 3.62	2.39 3.51	2.35 3.43	2.31 3.34	2.27 3.26	2.24 3.21	2.21 3.14	2.19 3.11	2.16 3.06	2.14 3.02	2.13 3.00	14
15	4.54 8.68	3.68 6.36	3.29 5.42	3.06 4.89	2.90 4.56	2.79 4.32	2.70 4.14	2.64 4.00	2.59 3.89	2.55 3.80	2.51 3.73	2.48 3.67	2.43 3.56	2.39 3.48	2.33 3.36	2.29 3.29	2.25 3.20	2.21 3.12	2.18 3.07	2.15 3.00	2.12 2.97	2.10 2.92	2.08 2.89	2.07 2.87	15
16	4.49 8.53	3.63 6.23	3.24 5.29	3.01 4.77	2.85 4.44	2.74 4.20	2.66 4.03	2.59 3.89	2.54 3.78	2.49 3.69	2.45 3.61	2.42 3.55	2.37 3.45	2.33 3.37	2.28 3.25	2.24 3.18	2.20 3.10	2.16 3.01	2.13 2.96	2.09 2.98	2.07 2.86	2.04 2.80	2.02 2.77	2.01 2.75	16
17	4.45 8.40	3.59 6.11	3.20 5.18	2.96 4.67	2.81 4.34	2.70 4.10	2.62 3.93	2.55 3.79	2.50 3.68	2.45 3.59	2.41 3.52	2.38 3.45	2.33 3.35	2.29 3.27	2.23 3.16	2.19 3.08	2.15 3.00	2.11 2.92	2.08 2.86	2.04 2.79	2.02 2.76	1.99 2.70	1.97 2.67	1.96 2.65	17
18	4.41 8.28	3.55 6.01	3.16 5.09	2.93 4.58	2.77 4.25	3.66 4.01	2.58 3.85	2.51 3.71	2.46 3.60	2.41 3.51	2.37 3.44	2.34 3.37	2.29 3.27	2.25 3.19	2.19 3.07	2.15 3.00	2.11 2.91	2.07 2.83	2.04 2.78	2.00 2.71	1.98 2.68	1.95 2.62	1.93 2.59	1.92 2.57	18
19	4.38 8.18	3.52 5.93	3.13 5.01	2.90 4.50	2.74 4.17	2.63 3.94	2.55 3.77	2.48 3.63	2.43 3.52	2.38 3.43	2.34 3.36	2.31 3.30	2.26 3.19	2.21 3.12	2.15 3.00	2.11 2.92	2.07 2.84	2.02 2.76	2.00 2.70	1.96 2.63	1.94 2.60	1.91 2.54	1.90 2.51	1.88 2.49	19
20	4.35 8.10	3.49 5.85	3.10 4.94	2.87 4.43	2.71 4.10	2.60 3.87	2.52 3.71	2.45 3.56	2.40 3.45	2.35 3.37	2.31 3.30	2.28 3.23	2.23 3.13	2.18 3.05	2.12 2.94	2.08 2.86	2.04 2.77	1.99 2.69	1.96 2.63	1.92 2.56	1.90 2.53	1.87 2.47	1.85 2.44	1.84 2.42	20
21	4.32 8.02	3.47 5.78	3.07 4.87	2.84 4.37	2.68 4.04	2.57 3.81	2.49 3.65	2.42 3.51	2.37 3.40	2.32 3.31	2.28 3.24	2.25 3.17	2.20 3.07	2.15 2.99	2.09 2.88	2.05 2.80	2.00 2.72	1.96 2.63	1.93 2.58	1.89 2.51	1.87 2.47	1.84 2.42	1.82 2.38	1.81 2.36	21
22	4.30 7.94	3.44 5.72	3.05 4.82	2.82 4.31	2.66 3.99	2.55 3.76	2.47 3.59	2.40 3.45	2.35 3.35	2.30 3.26	2.26 3.18	2.23 3.12	2.18 3.02	2.13 2.94	2.07 2.83	2.03 2.75	1.98 2.67	1.93 2.58	1.91 2.53	1.87 2.46	1.84 2.42	1.81 2.37	1.80 2.33	1.78 2.31	22
23	4.28 7.88	3.42 5.66	3.03 4.76	2.80 4.26	2.64 3.94	2.53 3.71	2.45 3.54	2.38 3.41	2.32 3.30	2.28 3.21	2.24 3.14	2.20 3.07	2.14 2.97	2.10 2.89	2.04 2.78	2.00 2.70	1.96 2.62	1.91 2.53	1.88 2.48	1.84 2.41	1.82 2.37	1.79 2.32	1.77 2.28	1.76 2.26	23
24	4.26 7.82	3.40 5.61	3.01 4.72	2.78 4.22	2.62 3.90	2.51 3.67	2.43 3.50	2.36 3.36	2.30 3.25	2.26 3.17	2.22 3.09	2.18 3.03	2.13 2.93	2.09 2.85	2.02 2.74	1.98 2.66	1.94 2.58	1.89 2.49	1.86 2.44	1.82 2.36	1.80 2.33	1.76 2.27	1.74 2.23	1.73 2.21	24
25	4.24 7.77	3.38 5.57	2.99 4.68	2.76 4.18	2.60 3.86	2.49 3.63	2.41 3.46	2.34 3.32	2.28 3.21	2.24 3.13	2.20 3.05	2.16 2.99	2.11 2.89	2.06 2.81	2.00 2.70	1.96 2.62	1.92 2.54	1.87 2.45	1.84 2.40	1.80 2.32	1.77 2.29	1.74 2.23	1.72 2.19	1.71 2.17	25
26	4.22 7.72	3.37 5.53	2.98 4.64	2.74 4.14	2.59 3.82	2.47 3.59	2.39 3.42	2.32 3.29	2.27 3.17	2.22 3.09	2.18 3.02	2.15 2.96	2.10 2.86	2.05 2.77	1.99 2.66	1.95 2.58	1.90 2.50	1.85 2.41	1.82 2.36	1.78 2.28	1.76 2.25	1.72 2.19	1.70 2.15	1.69 2.13	26

F TABLE—(Continued)

f_2	\multicolumn{17}{c}{f_1 Degrees of Freedom (for greater mean square)}	f_2
	1	2	3	4	5	6	7	8	9	10	11	12	14	16	20	24	30	40	50	75	100	200	500	∞	
27	4.21 7.68	3.35 5.49	2.96 4.60	2.73 4.11	2.57 3.79	2.46 3.56	2.37 3.39	2.30 3.26	2.25 3.14	2.20 3.06	2.16 2.98	2.13 2.93	2.08 2.83	2.03 2.74	1.97 2.63	1.93 2.55	1.88 2.47	1.84 2.38	1.80 2.33	1.76 2.25	1.74 2.21	1.71 2.16	1.68 2.12	1.67 2.10	27
28	4.20 7.64	3.34 5.45	2.95 4.57	2.71 4.07	2.56 3.76	2.44 3.53	2.36 3.36	2.29 3.23	2.24 3.11	2.19 3.03	2.15 2.95	2.12 2.90	2.06 2.80	2.02 2.71	1.96 2.60	1.91 2.52	1.87 2.44	1.81 2.35	1.78 2.30	1.75 2.22	1.72 2.18	1.69 2.13	1.67 2.09	1.65 2.06	28
29	4.18 7.60	3.33 5.42	2.93 4.54	2.70 4.04	2.54 3.73	2.43 3.50	2.35 3.33	2.28 3.20	2.22 3.08	2.18 3.00	2.14 2.92	2.10 2.87	2.05 2.77	2.00 2.68	1.94 2.57	1.90 2.49	1.85 2.41	1.80 2.32	1.77 2.27	1.73 2.19	1.71 2.15	1.68 2.10	1.65 2.06	1.64 2.03	29
30	4.17 7.56	3.32 5.39	2.92 4.51	2.69 4.02	2.53 3.70	2.42 3.47	2.34 3.30	2.27 3.17	2.21 3.06	2.16 2.98	2.12 2.90	2.09 2.84	2.04 2.74	1.99 2.66	1.93 2.55	1.89 2.47	1.84 2.38	1.79 2.29	1.76 2.24	1.72 2.16	1.69 2.13	1.66 2.07	1.64 2.03	1.62 2.01	30
32	4.15 7.50	3.30 5.34	2.90 4.46	2.67 3.97	2.51 3.66	2.40 3.42	2.32 3.25	2.25 3.12	2.19 3.01	2.14 2.94	2.10 2.86	2.07 2.80	2.02 2.70	1.97 2.62	1.91 2.51	1.86 2.42	1.82 2.34	1.76 2.25	1.74 2.20	1.69 2.12	1.67 2.08	1.64 2.02	1.61 1.98	1.59 1.96	32
34	4.13 7.44	3.28 5.29	2.88 4.42	2.65 3.93	2.49 3.61	2.38 3.38	2.30 3.21	2.23 3.08	2.17 2.97	2.12 2.89	2.08 2.82	2.05 2.76	2.00 2.66	1.95 2.58	1.89 2.47	1.84 2.38	1.80 2.30	1.74 2.21	1.71 2.15	1.67 2.08	1.64 2.04	1.61 1.98	1.59 1.94	1.57 1.91	34
36	4.11 7.39	3.26 5.25	2.86 4.38	2.63 3.89	2.48 3.58	2.36 3.35	2.28 3.18	2.21 3.04	2.15 2.94	2.10 2.86	2.06 2.78	2.03 2.72	1.98 2.62	1.93 2.54	1.87 2.43	1.82 2.35	1.78 2.26	1.72 2.17	1.69 2.12	1.65 2.04	1.62 2.00	1.59 1.94	1.56 1.90	1.55 1.87	36
38	4.10 7.35	3.25 5.21	2.85 4.34	2.62 3.86	2.46 3.54	2.35 3.32	2.26 3.15	2.19 3.02	2.14 2.91	2.09 2.82	2.05 2.75	2.02 2.69	1.96 2.59	1.92 2.51	1.85 2.40	1.80 2.32	1.76 2.22	1.71 2.14	1.67 2.08	1.63 2.00	1.60 1.97	1.57 1.90	1.54 1.86	1.53 1.84	38
40	4.08 7.31	3.23 5.18	2.84 4.31	2.61 3.83	2.45 3.51	2.34 3.29	2.25 3.12	2.18 2.99	2.12 2.88	2.07 2.80	2.04 2.73	2.00 2.66	1.95 2.56	1.90 2.49	1.84 2.37	1.79 2.29	1.74 2.20	1.69 2.11	1.66 2.05	1.61 1.97	1.59 1.94	1.55 1.88	1.53 1.84	1.51 1.81	40
42	4.07 7.27	3.22 5.15	2.83 4.29	2.59 3.80	2.44 3.49	2.32 3.26	2.24 3.10	2.17 2.96	2.11 2.86	2.06 2.77	2.02 2.70	1.99 2.64	1.94 2.54	1.89 2.46	1.82 2.35	1.78 2.26	1.73 2.17	1.68 2.08	1.64 2.02	1.60 1.94	1.57 1.91	1.54 1.85	1.51 1.80	1.49 1.78	42
44	4.06 7.24	3.21 5.12	2.82 4.26	2.58 3.78	2.43 3.46	2.31 3.24	2.23 3.07	2.16 2.94	2.10 2.84	2.05 2.75	2.01 2.68	1.98 2.62	1.92 2.52	1.88 2.44	1.81 2.32	1.76 2.24	1.72 2.15	1.66 2.06	1.63 2.00	1.58 1.92	1.56 1.88	1.52 1.82	1.50 1.78	1.48 1.75	44
46	4.05 7.21	3.20 5.10	2.81 4.24	2.57 3.76	2.42 3.44	2.30 3.22	2.22 3.05	2.14 2.92	2.09 2.82	2.04 2.73	2.00 2.66	1.97 2.60	1.91 2.50	1.87 2.42	1.80 2.30	1.75 2.22	1.71 2.13	1.65 2.04	1.62 1.98	1.57 1.90	1.54 1.86	1.51 1.80	1.48 1.76	1.46 1.72	46
48	4.04 7.19	3.19 5.08	2.80 4.22	2.56 3.74	2.41 3.42	2.30 3.20	2.21 3.04	2.14 2.90	2.08 2.80	2.03 2.71	1.99 2.64	1.96 2.58	1.90 2.48	1.86 2.40	1.79 2.28	1.74 2.20	1.70 2.11	1.64 2.02	1.61 1.96	1.56 1.88	1.53 1.84	1.50 1.78	1.47 1.73	1.45 1.70	48

F TABLE—(Continued)

| f_2 | \multicolumn{24}{c|}{f_1, Degrees of Freedom (for greater mean square)} | f_2 |

f_2	1	2	3	4	5	6	7	8	9	10	11	12	14	16	20	24	30	40	50	75	100	200	500	∞	f_2
50	4.03/7.17	3.18/5.06	2.79/4.20	2.56/3.72	2.40/3.41	2.29/3.18	2.20/3.02	2.13/2.88	2.07/2.78	2.02/2.70	1.98/2.62	1.95/2.56	1.90/2.46	1.85/2.39	1.78/2.26	1.74/2.18	1.69/2.10	1.63/2.00	1.60/1.94	1.55/1.86	1.52/1.82	1.48/1.76	1.46/1.71	1.44/1.68	50
55	4.02/7.12	3.17/5.01	2.78/4.16	2.54/3.68	2.38/3.37	2.27/3.15	2.18/2.98	2.11/2.85	2.05/2.75	2.00/2.66	1.97/2.59	1.93/2.53	1.88/2.43	1.83/2.35	1.76/2.23	1.72/2.15	1.67/2.06	1.61/1.96	1.58/1.90	1.52/1.82	1.50/1.78	1.46/1.71	1.43/1.66	1.41/1.64	55
60	4.00/7.08	3.15/4.98	2.76/4.13	2.52/3.65	2.37/3.34	2.25/3.12	2.17/2.95	2.10/2.82	2.04/2.72	1.99/2.63	1.95/2.56	1.92/2.50	1.86/2.40	1.81/2.32	1.75/2.20	1.70/2.12	1.65/2.03	1.59/1.93	1.56/1.87	1.50/1.79	1.48/1.74	1.44/1.68	1.41/1.63	1.39/1.60	60
65	3.99/7.04	3.14/4.95	2.75/4.10	2.51/3.62	2.36/3.31	2.24/3.09	2.15/2.93	2.08/2.79	2.02/2.70	1.98/2.61	1.94/2.54	1.90/2.47	1.85/2.37	1.80/2.30	1.73/2.18	1.68/2.09	1.63/2.00	1.57/1.90	1.54/1.84	1.49/1.76	1.46/1.71	1.42/1.64	1.39/1.60	1.37/1.56	65
70	3.98/7.01	3.13/4.92	2.74/4.08	2.50/3.60	2.35/3.29	2.23/3.07	2.14/2.91	2.07/2.77	2.01/2.67	1.97/2.59	1.93/2.51	1.89/2.45	1.84/2.35	1.79/2.28	1.72/2.15	1.67/2.07	1.62/1.98	1.56/1.88	1.53/1.82	1.47/1.74	1.45/1.69	1.40/1.62	1.37/1.56	1.35/1.53	70
80	3.96/6.96	3.11/4.88	2.72/4.04	2.48/3.56	2.33/3.25	2.21/3.04	2.12/2.87	2.05/2.74	1.99/2.64	1.95/2.55	1.91/2.48	1.88/2.41	1.82/2.32	1.77/2.24	1.70/2.11	1.65/2.03	1.60/1.94	1.54/1.84	1.51/1.78	1.45/1.70	1.42/1.65	1.38/1.57	1.35/1.52	1.32/1.49	80
100	3.94/6.90	3.09/4.82	2.70/3.98	2.46/3.51	2.30/3.20	2.19/2.99	2.10/2.82	2.03/2.69	1.97/2.59	1.92/2.51	1.88/2.43	1.85/2.36	1.79/2.26	1.75/2.19	1.68/2.06	1.63/1.98	1.57/1.89	1.51/1.79	1.48/1.73	1.42/1.64	1.39/1.59	1.34/1.51	1.30/1.46	1.28/1.43	100
125	3.92/6.84	3.07/4.78	2.68/3.94	2.44/3.47	2.29/3.17	2.17/2.95	2.08/2.79	2.01/2.65	1.95/2.56	1.90/2.47	1.86/2.40	1.83/2.33	1.77/2.23	1.72/2.15	1.65/2.03	1.60/1.94	1.55/1.85	1.49/1.75	1.45/1.68	1.39/1.59	1.36/1.54	1.31/1.46	1.27/1.40	1.25/1.37	125
150	3.91/6.81	3.06/4.75	2.67/3.91	2.43/3.44	2.27/3.14	2.16/2.92	2.07/2.76	2.00/2.62	1.94/2.53	1.89/2.44	1.85/2.37	1.82/2.30	1.76/2.20	1.71/2.12	1.64/2.00	1.59/1.91	1.54/1.83	1.47/1.72	1.44/1.66	1.37/1.56	1.34/1.51	1.29/1.43	1.25/1.37	1.22/1.33	150
200	3.89/6.76	3.04/4.71	2.65/3.88	2.41/3.41	2.26/3.11	2.14/2.90	2.05/2.73	1.98/2.60	1.92/2.50	1.87/2.41	1.83/2.34	1.80/2.28	1.74/2.17	1.69/2.09	1.62/1.97	1.57/1.88	1.52/1.79	1.45/1.69	1.42/1.62	1.35/1.53	1.32/1.48	1.26/1.39	1.22/1.33	1.19/1.28	200
400	3.86/6.70	3.02/4.66	2.62/3.83	2.39/3.36	2.23/3.06	2.12/2.85	2.03/2.69	1.96/2.55	1.90/2.46	1.85/2.37	1.81/2.29	1.78/2.23	1.72/2.12	1.67/2.04	1.60/1.92	1.54/1.84	1.49/1.74	1.42/1.64	1.38/1.57	1.32/1.47	1.28/1.42	1.22/1.32	1.16/1.24	1.13/1.19	400
1000	3.85/6.66	3.00/4.62	2.61/3.80	2.38/3.34	2.22/3.04	2.10/2.82	2.02/2.66	1.95/2.53	1.89/2.43	1.84/2.34	1.80/2.26	1.76/2.20	1.70/2.09	1.65/2.01	1.58/1.89	1.53/1.81	1.47/1.71	1.41/1.61	1.36/1.54	1.30/1.44	1.26/1.38	1.19/1.28	1.13/1.19	1.08/1.11	1000
∞	3.84/6.64	2.99/4.60	2.60/3.78	2.37/3.32	2.21/3.02	2.09/2.80	2.01/2.64	1.94/2.51	1.88/2.41	1.83/2.32	1.79/2.24	1.75/2.18	1.69/2.07	1.64/1.99	1.57/1.87	1.52/1.79	1.46/1.69	1.40/1.59	1.35/1.52	1.28/1.41	1.24/1.36	1.17/1.25	1.11/1.15	1.00/1.00	∞

Appendix B: CALSTAT

By using the accompanying CD, you accept the terms of the CALSTAT License Agreement included in this book.

CALSTAT License Agreement

Oxford University Press and David J. Weiss grant to the purchaser of this book a nonexclusive license to install and use the software on no more than two computers. Copyright to the individual programs recorded on the disk is owned by the author, David J. Weiss.

You may make a copy of the software for backup purposes only. You may not reverse engineer, decompile, or disassemble the software. You may transfer the software on a permanent basis, provided that the transferee agrees to accept the terms and conditions of this Agreement and you retain no copies. You may not rent or lease the software.

The software is provided "as is" without warranty of any kind, either express or implied from Oxford or David J. Weiss. Any risk arising out of use of the software remains with the user.

Installation Instructions for CALSTAT

The CALSTAT package installs through a standard Windows setup program. Close other applications, then insert the CD into the appropriate drive (we'll call it e:). From the Start menu, click on Run . . . Type in e:\Setup (use the correct drive name for your system). Alternatively, you may double-click on Setup.exe (*not* Setup.Lst) from Windows Explorer (having opened drive e:).

By default, Setup will offer to install the programs in the C:\Program Files\Calstat directory; you may choose another location, but there probably is no reason to do so. Icons will be automatically created in the CALSTAT group.

Required Software and Hardware

The program requires a personal computer with a CD drive, running Microsoft Windows 95 or later.

Terms from Introductory Statistics

A priori: Latin phrase for designated in advance. The usual statistical hypotheses are regarded as having been specified without reference to the data. Tests are carried out as though the hypotheses had been proposed in advance even if the researcher had been lax about specification. The alternative to prior designation is a post-hoc test, so named because the hypothesis to be evaluated is constructed as a result of something interesting having emerged in the data.

Alternative hypothesis: In general, the proposition expressing the particular way the null hypothesis is held to be false. Sometimes referred to as the motivated hypothesis, it usually reflects a difference the researcher hopes to demonstrate. In the ANOVA setting, the usual alternative hypothesis is that the true means of the various groups are unequal.

Confidence interval: A symmetrical region constructed around the sample mean; the region has a specified probability of containing the true mean. The 95% confidence interval has a .95 probability of containing the true mean, while the correspondingly larger 99% confidence interval has a .99 probability of containing the true mean.

Confounded: The situation in which the effect of a controlled variable is inextricably mixed with that of another, uncontrolled variable. For example, if we select children from a schoolyard and form experimental groups according to their heights, the groups would be very likely to differ in age as well.

Critical value: A value taken from a statistical table. It serves as the criterion for determining whether the corresponding data-based statistic is large enough to be considered evidence against the null hypothesis.

Degrees of freedom (df): A structural aspect of the experimental design, determined by the number of scores and not their values. The general rule is that each observation generates one *df*. The number of elements in each subdivision of the design determines the allocation of *df*s to structural units. Some statistical tests, among them F and t, require *df* specification to determine critical values.

Dependent variable: An index of the behavior of interest. In univariate analysis (ANOVA, the subject of this book), there is one behavior selected by the researcher to reflect an interesting psychological property. In MANOVA (multivariate analysis of variance), there is more than one dependent variable.

Group: A collection of participants given the same experimental treatment. The group is formed within the research project and need not have existed prior to the experiment. Participants in a group may actually generate their data on an individual basis and never come into contact with one another. It is their scores that are grouped for analytic purposes.

Independent variable: A treatment under experimental control. The researcher imposes several (at least two) different values of the independent variable on participants, and compares the results to determine whether the variable differentially affects behavior. An experiment must have at least one independent variable, and may have several.

Interval scale: A measuring system having the property that intervals between the numbers assigned accurately reflect intervals between the magnitudes of the objects being measured. The important implication is that a given difference between two scores is independent of location along the scale.

Invalid: When applied to a measurement scale, a lack of validity means that the numbers assigned do not reflect the true magnitude of the property being measured. (Pronounced with the second syllable accented, to distinguish this adjective from the homographic noun designating a person in poor health.)

Nominal scale: A measuring system in which the numbers used do not reflect the magnitude of the objects being measured. The numbers serve only to label or classify, not to order in any way.

Null hypothesis: In the ANOVA context, a statement that there is no difference among a set of true means. In other statistical settings, the statement similarly refers to an absence of interesting differences. Because observed means are estimates of the true values, statistical machinery is invoked to determine the validity of the hypothesis. The fallibility of data implies that some incorrect decisions are inevitable.

One-tailed test: A significance test in which a statistic obtained from the data is compared to the critical value in a designated tail of the probability distribution. Used when the alternative hypothesis specifies the sign the test statistic must achieve. Synonymous with directional hypothesis. Because F ratios must be positive, ANOVA employs only one-tailed tests.

Ordinal scale: A measuring system in which the numbers used are monotonically related to the magnitude of the objects being measured. Monotonicity means the larger the object, the larger the number.

Power: The ability of a statistical test to verify the falsity of the null hypothesis. Power is expressed as $1 - \beta$, where β is the probability of a Type II error. In general, the specific value of β is unknown, but it is affected by many aspects of the experiment as well as by the statistical procedures chosen by the researcher. Obviously, having a powerful experiment is desirable.

Probability distribution: The possible outcomes of an experiment along with their associated probabilities. Specific probability distributions, such as the normal, t, and F have been derived from sets of assumptions about how scores are generated and the way they are combined. When the assumptions are correct, the probability distribution may be used to determine the critical value for a significance test. In practice, the distribution is presumed applicable given the structure of the experiment and the assumptions are not specifically checked.

***p* value:** The probability associated with an obtained statistic, such as an F ratio. If the *p* value is less than the significance level, then the result constitutes evidence against the null hypothesis.

Ratio scale: A measuring system having the property that ratios of the numbers assigned accurately reflect ratios of the magnitudes of the objects being measured. This means that a behavior assigned a number of, say, 10, must have twice the magnitude of a behavior assigned a number of 5. Effectively, this requires that there be a true zero, a score that means the behavior is absent. Ratio scales in psychology are almost always physical scales employed to capture behavior, such as the use of response latency to measure task difficulty. Typical "psychological" dimensions, for example, intelligence or attitude, do not have a "zero" point.

Significance level: A value selected by the researcher to fix the probability of a Type I error. Denoted as α, this arbitrary choice is always a small probability. The most common choice in behavioral research is .05, while .01 is occasionally used. The significance level determines the critical value drawn from the statistical table.

Significant difference: A difference among observed values, such as group means, deemed large enough to be considered reliable. Determination of significance comes from a statistical test, and carries with it the probabilistic characterization of significance level. As used in the statistics context, the word does not have its everyday connotation of importance.

Subscript notation:

$$X_1, X_2, X_3, \ldots, X_n$$

The subscripts attached to each of the Xs in a set offer a handy way to refer to any of them or to all n of them at once. As an example, suppose we have measured the weights of five objects: 12 grams, 8 grams, 5 grams, 10 grams, and 8 grams. If we refer to the weights as Xs, we would say $X_1 = 12$, $X_2 = 8$, and so on. n would be five, since there are five scores. What is the sum of $X_3 + X_4$? 15 grams.

Substantive variable: An element of theoretical interest within the experiment. The researcher plans to determine whether different values of the substantive variable produce different responses.

Summation: The Greek capital letter sigma, Σ, is used as a shorthand way of indicating the sum of a set of numbers expressed with *subscript notation*. It sometimes appears with an index and with limits, for example,

$$\sum_{i=1}^{i=4} i$$

This is read as "the sum of the *i*s, where *i* goes from 1 to 4"; thus the sum is 10. More frequently, the index appears as a subscript. Using the weights we measured for our discussion of subscript notation,

$$\sum_{i=1}^{i=4} X_i$$

When no index or limits are furnished, the summation is considered to extend over all of the scores under discussion; so $\Sigma X = 43$ grams.

Two-tailed test: A significance test in which two critical regions are defined, one in each tail of the probability distribution. Significance is achieved when the observed value of the test statistic is more extreme than either critical value. A two-tailed test is used when the sign of the test statistic is not specified by the alternative hypothesis. Thus a two-tailed test would be used if the alternative hypothesis were that two quantities are unequal, without prior specification of which is larger. Synonymous with nondirectional test.

Type I error: The researcher's data-based decision that the null hypothesis is false when it is really true. This incorrect conclusion is not the result of a mistake in the analysis. By chance, a "large" dispersion among the means (which should happen with probability α) has actually occurred this time.

Type II error: The researcher's data-based decision that the null hypothesis is true when it is really false. While this incorrect conclusion could be the result of bad luck as in the case of a Type I error, a more worrisome possibility is that the experiment had low power.

Variance: Variance is a formal measure of the dispersion, or scatter, among the numbers in a set of data. The variance is a weighted average of the differences between the numbers, with larger discrepancies contributing more heavily. Other measures of dispersion are also plausible, but the variance is the most popular because of its "mathematical tractability" and its central role in statistical theory. It is convenient to define variance in terms of the discrepancies from the mean ($\bar{x}$); basically, this is equivalent to an alternative definition considering the differences between all pairs of numbers, but is much simpler computationally. Thus the definition:

$$\text{Variance} = \frac{\Sigma(x_i - \bar{x})^2}{n-1}$$

This definition may be seen to sum the deviations from the mean; squaring each deviation accomplishes the weighting such that large deviations contribute very heavily to the summation. Dividing the sum by $n - 1$ accomplishes the averaging. Division by $n - 1$ rather than by n allows the variance in a sample to be an unbiased estimator of the variance in the population (further discussion of estimators is unnecessary for our purposes, but may be found in an intermediate-level text on mathematical statistics). Since data always should be regarded as sample values, this is the appropriate definition in an empirical field. In the world of ANOVA, the term "mean square" is used as a functional equivalent of variance.

References

Anderson, N. H. (1961). Scales and statistics: Parametric and nonparametric. *Psychological Bulletin, 58,* 305–316.

Anderson, N. H. (1968). Partial analysis of high-way factorial designs. *Behavior Methods and Instrumentation, 1,* 2–7.

Anderson, N. H. (1970). Functional measurement and psychophysical judgment. *Psychological Review, 77,* 153–170.

Anderson, N. H. (1976). How functional measurement can yield validated interval scales of mental quantities. *Journal of Applied Psychology, 61,* 677–692.

Anderson, N. H. (1977). Failure of additivity in bisection of length. *Perception & Psychophysics, 22,* 213–222.

Anderson, N. H. (1981). *Foundations of information integration theory.* New York: Academic Press.

Anderson, N. H. (1982). *Methods of information integration theory.* New York: Academic Press.

Anderson, N. H. (1990). Personal design in social cognition. In C. Hendrick & M. S. Clark (Eds.), *Research methods in personality and social psychology: Review of personality and social psychology* (Vol. 11, pp. 243–278). Beverly Hills, CA: Sage.

Anderson, N. H. (2001). *Empirical direction in design and analysis.* Mahwah, NJ: Lawrence Erlbaum Associates.

Anderson, N. H., & Armstrong, M. A. (1989). Cognitive theory and methodology for studying marital interaction. In D. Brinberg & J. Jaccard (Eds.), *Dyadic decision making* (pp. 3–50). New York: Springer-Verlag.

Anderson, N. H., & Shanteau, J. C. (1970). Information integration in risky decision making. *Journal of Experimental Psychology, 84,* 441–451.

Anderson, N. H., & Shanteau, J. C. (1977). Weak inference with linear models. *Psychological Bulletin, 84,* 1155–1170.

Anderson, N. H., & Weiss, D. J. (1971). Test of a multiplying model for estimated area of rectangles. *American Journal of Psychology, 84,* 543–548.

Bernhardson, C. S. (1975). Type I error rates when multiple comparison procedures follow a significant F test of ANOVA. *Biometrics, 31*, 229–232.

Birnbaum, M. H., & Veit, C. T. (1974). Scale convergence as a criterion for rescaling: Information integration with difference, ratio, and averaging tasks. *Perception & Psychophysics, 15*, 7–15.

Busemeyer, J. R. (1980). Importance of measurement theory, error theory, and experimental design for testing the significance of interactions. *Psychological Bulletin, 88*, 237–244.

Cochran, W. G. (1943). The comparison of different scales of measurement for experimental results. *Annals of Mathematical Statistics, 14*, 205–216.

Cochran, W. G., & Cox, G. M. (1957). *Experimental designs* (2nd ed.). New York: Wiley.

Cohen, J. (1968). Multiple regression as a general data analytic system. *Psychological Bulletin, 70*, 426–443.

Cohen, J. (1977). *Statistical power analysis for the behavioral sciences* (2nd ed.). Hillsdale, NJ: Erlbaum.

Cronbach, L. J., & Furby, L. (1970). How we should measure "change"—or should we? *Psychological Bulletin, 74*, 68–80.

Davis, C., & Gaito, J. (1984). Multiple comparison procedures within experimental research. *Canadian Psychology/Psychologie Canadienne, 25*, 1–13.

Davison, M., & Sharma, A. (1988). Parametric statistics and levels of measurement. *Psychological Bulletin, 104*, 137–144.

Elder, W. W., & Weiss, D. J. (1987). SNAPSHOT: Analysis of variance with unequal numbers of scores per subject. *Educational and Psychological Measurement, 47*, 117–119.

Emerson, P. L. (1965). A FORTRAN generator of polynomials orthonormal over unequally spaced and weighted abscissa. *Educational and Psychological Measurement, 25*, 867–871.

Feldt, L. S. (1958). A comparison of the precision of three experimental designs employing a concomitant variable. *Psychometrika, 23*, 335–353.

Fisher, R. A. (1935). *Statistical methods for research workers.* Edinburgh: Oliver & Boyd.

Gaito, J. (1978). Multiple comparisons within ANOVA using orthogonal and nonorthogonal components. *Educational and Psychological Measurement, 38*, 901–904.

Gaito, J. (1980). Measurement scales and statistics: Resurgence of an old misconception. *Psychological Bulletin, 87*, 564–567.

Glass, G. V., & Hakstian, A. R. (1969). Measures of association in comparative experiments: Their development and interpretation. *American Educational Research Journal, 6*, 403–414.

Grice, G. R. (1966). Dependence of empirical laws upon the source of experimental variation. *Psychological Bulletin, 66*, 488–498.

Haralson, J. V., & Richert, M. (1979). Adjunctive aggression as a function of food deprivation and habituation in the fish *Serotherdon melanotheron*. Unpublished manuscript, California State University, Los Angeles.

Hays, W. L. (1963). *Statistics for psychologists.* New York: Holt, Rinehart, and Winston.

Herr, D. G., & Gabelein, J. (1978). Nonorthogonal two-way analysis of variance. *Psychological Bulletin, 85*, 207–216.

Hoenig, J. M., & Heisey, D. M. (2001). The abuse of power: The pervasive fallacy of power calculations for data analysis. *The American Statistician, 55*, 19–24.

Howe, E. S. (1991). Integration of mitigation, intention, and outcome damage information by students and circuit court judges. *Journal of Applied Social Psychology, 21*, 875–895.

Huck, S. W., & McLean, R. A. (1975). Using a repeated measures ANOVA to analyze the data from a pretest-posttest design: A potentially confusing task. *Psychological Bulletin, 82*, 511–518.

Hurlbut, R. T., & Spiegel, D. K. (1976). Dependence of F-ratios sharing a common denominator mean square. *American Statistician, 30*, 74–76.
Jaccard J. J., & Becker, M. A. (1985). Attitudes and behavior: An information integration perspective. *Journal of Experimental Social Psychology, 21*, 440–465.
Jacobson, N. S., Roberts, L. J., Berns, S. B., & McGlinchey, J. B. (1999). Methods for defining and determining the clinical significance of treatment effects: Description, application, and alternatives. *Journal of Consulting and Clinical Psychology, 67*, 300–307.
Keppel, G. (1991). *Design and analysis: A researcher's handbook* (3rd ed.). Upper Saddle River, NJ: Prentice-Hall.
Keren, G., & Lewis, C. (1979). Partial omega squared for ANOVA designs. *Educational and Psychological Measurement, 39*, 119–128.
Krantz, D. H., Luce, R. D., Suppes, P., & Tversky, A. (1971). *Foundations of measurement* (Vol. 1). New York: Academic Press.
Krueger, L. E. (1989). Reconciling Fechner and Stevens: Toward a unified psychophysical law. *Behavioral and Brain Sciences, 12*, 251–320.
Kruskal, J. B. (1965). Analysis of factorial experiments by estimating monotone transformations of the data. *Journal of the Royal Statistical Society (Series B), 27*, 251–263.
Lenth, R. V. (2001). Some practical guidelines for effective sample size determination. *The American Statistician, 55*, 187–193.
Little, R. J. A., & Rubin, D. B. (1987). *Statistical analysis with missing data*. New York: Wiley.
Lord, F. M. (1953). On the statistical treatment of football numbers. *American Psychologist, 8*, 750–751.
Luce, R. D., & Tukey, J. W. (1964). Simultaneous conjoint measurement: A new type of fundamental measurement. *Journal of Mathematical Psychology, 1*, 1–27.
Maxwell, S. E. (2004). The persistence of underpowered studies in psychological research: Causes, consequences, and remedies. *Psychological Methods, 9*, 147–163.
Maxwell, S. E., Camp, C. J., & Arvey, R. D. (1981). Measures of strength of association: A comparative examination. *Journal of Applied Psychology, 66*, 525–534.
Maxwell, S. E., Delaney, H. D., & Dill, C. A. (1984). Another look at ANCOVA versus blocking. *Psychological Bulletin, 95*, 136–147.
Maxwell, S. E., & Howard, G. S. (1981). Change scores—necessarily anathema? *Educational and Psychological Measurement, 41*, 747–756.
Milligan, G. W., Wong, D. S., & Thompson, P. A. (1987). Robustness properties of nonorthogonal analysis of variance. *Psychological Bulletin, 101*, 464–470.
O'Grady, K. E. (1982). Measures of explained variance: Cautions and limitations. *Psychological Bulletin, 92*, 766–777.
Pearson, E. S., & Hartley, H. O. (1951). Charts of the power function for analysis of variance tests, derived from the non-central F distribution. *Biometrika, 38*, 112–130.
Petrinovich, L. F., & Hardyck, C. D. (1969). Error rates for multiple comparison methods: Some evidence concerning the frequency of erroneous conclusions. *Psychological Bulletin, 71*, 43–54.
Pimsleur, P., & Bonkowski, R. (1961). Transfer of verbal material across sense modalities. *Journal of Educational Psychology, 52*, 104–107.
Poulton, E. C. (1973). Unwanted range effects from using within-subject experimental designs. *Psychological Bulletin, 80*, 113–121.
Rosenthal, R., & Gaito, J. (1963). The interpretation of significance by psychological researchers. *Journal of Psychology, 55*, 33–38.
Rundall, C. S., & Weiss, D. J. (1994). Nurses' fear of contagion: A functional measurement analysis. *Medical Decision Making, 14*, 40–45.
Rundall, C. S., & Weiss, D. J. (1998). Patients' anticipated compliance: A functional measurement analysis. *Psychology, Health & Medicine, 3*, 261–274.

Schafer, J. L. (1999). Multiple imputation: A primer. *Statistical Methods in Medical Research, 8*, 3–15.

Schafer, J. L., & Graham, J. W. (2002). Missing data: Our view of the state of the art. *Psychological Methods, 7*, 147–177.

Scheffé, H. A. (1953). A method for judging all possible contrasts in the analysis of variance. *Biometrika, 40*, 87–104.

Schumann, D. E. W., & Bradley, R. A. (1959). The comparison of the sensitivities of similar experiments: Model II of the analysis of variance. *Biometrics, 15*, 405–416.

Shanteau, J. C. (1970). An additive model for sequential decision making. *Journal of Experimental Psychology, 85*, 181–191.

Shanteau, J. C., & Anderson, N. H. (1969). Test of a conflict model for preference judgment. *Journal of Mathematical Psychology, 6*, 312–365.

Stevens, S. S. (1951). Mathematics, measurement, and psychophysics. In S. S. Stevens (Ed.), *Handbook of experimental psychology* (pp. 1–41). New York: Wiley.

Strube, M. J. (1988). A BASIC program for the generation of Latin squares. *Behavior Research Methods, Instruments, and Computers, 20*, 508–509.

Suppes, P., & Zinnes, J. L. (1963). Basic measurement theory. In R. D. Luce, R. R. Bush, & E. Galanter (Eds.), *Handbook of mathematical psychology, Vol. 1* (pp. 1–76). New York: Wiley.

Surber, C. F. (1984). Inferences of ability and effort: Evidence for two different processes. *Journal of Personality and Social Psychology, 46*, 249–268.

Toothaker, L. E. (1991). *Multiple comparisons for researchers.* Newbury Park, CA: Sage.

Townsend, J. T., & Ashby, F. G. (1984). Measurement scales and statistics: The misconception misconceived. *Psychological Bulletin, 96*, 394–401.

Tukey, J. W. (1953). *The problem of multiple comparisons.* Mimeographed monograph, Princeton University. (Note: this is a secondary citation. Many statistics texts cite this work, although the original monograph is no longer accessible.)

Vaughan, G. M., & Corballis, M. C. (1969). Beyond tests of significance: Estimating strength of effects in selected ANOVA designs. *Psychological Bulletin, 72*, 204–213.

Weiss, D. J. (1972). Averaging: An empirical validity criterion for magnitude estimation. *Perception and Psychophysics, 12*, 385–388.

Weiss, D. J. (1973). FUNPOT, a FORTRAN program for finding a polynomial transformation to reduce any source(s) of variance in a factorial design. *Behavioral Science, 18*, 150.

Weiss, D. J. (1975). Quantifying private events: A functional measurement analysis of equisection. *Perception and Psychophysics, 17*, 351–357.

Weiss, D. J. (1980a). Note on choosing a response scale. *Perceptual and Motor Skills, 50*, 472–474.

Weiss, D. J. (1980b). ORPOCO: Orthogonal polynomial coefficients. *Behavior Research Methods and Instrumentation, 12*, 635.

Weiss, D. J. (1982). REPCOP: Repeated-measures planned comparisons and orthogonal polynomials. *Behavior Research Methods and Instrumentation, 14*, 44.

Weiss, D. J. (1985a). Snapshot analysis of variance: Comparing groups with unequal numbers of scores per subject. *Perceptual and Motor Skills, 61*, 420–422.

Weiss, D. J. (1985b). SCHUBRAD: The comparison of the sensitivities of similar experiments. *Behavior Research Methods, Instrumentation, and Computers, 17*, 572.

Weiss, D. J. (1986). The discriminating power of ordinal data. *Journal of Social Behavior and Personality, 1*, 381–389.

Weiss, D. J. (1987). Dropout in patient compliance studies: A suggested analytic procedure. *Journal of Compliance in Health Care, 2*, 73–83.

Weiss, D. J. (1991). A behavioral assumption for the analysis of missing data: The use of implanted zeroes. *Journal of Social Behavior and Personality, 6*, 955–964.

Weiss, D. J. (1999). An analysis of variance test for random attrition. *Journal of Social Behavior and Personality, 14*, 433–438.

Weiss, D. J., & Anderson, N. H. (1969). Subjective averaging of length with serial presentation. *Journal of Experimental Psychology, 82*, 52–63.
Weiss, D. J., & Black, J. K. (1995). Cognitive components of the decision to rape. Paper presented at the Mathematical Psychology Meeting, Irvine, CA, August.
Weiss, D. J., Edwards, W., & Weiss, J. W. (2005). The clinical significance decision. Manuscript submitted for publication.
Weiss, D. J., & Gardner, G. S. (1979). Subjective hypotenuse estimation: A test of the Pythagorean theorem. *Perceptual and Motor Skills, 48*, 607–615.
Weiss, D. J., & Shanteau, J. C. (1982). Group-Individual POLYLIN. *Behavior Research Methods and Instrumentation, 14*, 430.
Weiss, D. J., & Shanteau, J. (2003). Empirical assessment of expertise. *Human Factors, 45*, 104–116.
Weiss, D. J., Walker, D. L., & Hill, D. (1988). The choice of a measure in a health-promotion study. *Health Education Research: Theory and Practice, 3*, 381–386.
Wilkinson, L., & APA Task Force on Statistical Inference. (1999). Statistical methods in psychology journals: Guidelines and explanations. *American Psychologist, 54*, 594–604.
Winer, B. J. (1971). *Statistical principles in experimental design*. New York: McGraw-Hill.
Young, F. W. (1972). A model for polynomial conjoint analysis algorithms. In R. W. Shepard, A. K. Romney, & S. B. Nerlove (Eds.), *Multidimensional scaling: Theory and applications in the behavioral sciences. Volume 1: Theory* (pp. 69–104). New York: Seminar Press.
Zalinski, J., & Anderson, N. H. (1986). AVERAGE, a user-friendly FORTRAN-77 program for parameter estimation for the averaging model of information integration theory. La Jolla, CA: University of California, San Diego.

Author Index

Anderson, N. H., 56, 71, 122, 127, 141n.4, 210–13, 215, 224, 228, 244nn.2–3
Armstrong, M. A., 211
Arvey, R. D., 132
Ashby, F. G., 122

Becker, M. A., 215
Bernhardson, C. S., 114
Berns, S. B., 141n.3
Birnbaum, M. H., 213
Black, J. K., 137
Bonkowski, R., 98n.4
Bradley, R. A., 134, 135
Busemeyer, J. R., 56

Camp, C. J., 132
Cochran, W. G., 134, 195, 196, 199
Cohen, J., 6, 130, 137
Corballis, M. C., 141n.1
Cox, G. M., 195, 196, 199
Cronbach, L. J., 153

Davis, C., 105
Davison, M., 123
Delaney, H. D., 98n.3
Dill, C. A., 98n.3

Edwards, W., 141n.3
Elder, W. W., 177, 188n.4
Emerson, P. L., 108

Feldt, L. S., 88
Fisher, R. A., 11, 133
Furby, L., 153

Gabelein, J., 188n.1
Gaito, J., 105, 120n.1, 122, 129
Gardner, G. S., 93, 244n.4
Glass, G. V., 132
Graham, J. W., 170
Grice, G. R., 83

Hakstian, A. R., 132
Haralson, J. V., 92
Hardyck, C. D., 105
Hartley, H. O., 137
Hays, W. L., 130
Heisey, D. M., 137
Herr, D. G., 188n.1
Hill, D., 63
Hoenig, J. M., 137
Howard, G. S., 153
Howe, E. S., 216

Huck, S. W., 153
Hurlbut, R. T., 72

Jaccard, J. J., 215
Jacobson, N. S., 141n.3

Keppel, G., 57, 98n.3, 137
Keren, G., 131, 132
Krantz, D. H., 244n.3
Krueger, L. E., 212
Kruskal, J. B., 125, 218

Lenth, R. V., 137
Lewis, C., 131, 132
Little, R. J. A., 170
Lord, F. M., 121
Luce, R. D., 244n.3

Maxwell, S. E., 98n.3, 130, 132, 153
McGlinchey, J. B., 141n.3
McLean, R. A., 153
Milligan, G. W., 188n.1

O'Grady, K. E., 132

Pearson, E. S., 137
Petrinovich, L. F., 105
Pimsleur, P., 98n.4
Poulton, E. C., 98n.1

Richert, M., 92
Roberts, L. J., 141n.3
Rosenthal, R., 129
Rubin, D. B., 170
Rundall, C. S., 215, 216, 229

Schafer, J. L., 170
Scheffé, H. A., 112
Schumann, D. E. W., 134, 135
Shanteau, J. C., 56, 58, 65, 210, 215, 224, 244n.1
Sharma, A., 123
Spiegel, D. K., 72
Stevens, S. S., 122
Strube, M. J., 195
Suppes, P., 122, 244n.3
Surber, C. F., 223

Thompson, P. A., 188n.1
Toothaker, L. E., 120n.2
Townsend, J. T., 122
Tukey, J. W., 114, 244n.3
Tversky, A., 244n.3

Vaughan, G. M., 141n.1
Veit, C. T., 213

Walker, D. L., 63
Weiss, D. J., 58, 63, 74, 93, 108, 123, 125, 135–37, 141n.2, 162, 173, 176, 177, 188n.3, 212–13, 215, 216, 218, 228, 229, 244nn.1, 4
Weiss, J. W., 141n.3
Wilkinson, L., 130
Winer, B. J., 169
Wong, D. S., 188n.1

Young, F. W., 125

Zalinski, J., 244n.2
Zinnes, J. L., 122

Subject Index

a priori transformation, defined, 247
accidental inequality, coping with, 163
adding model, 218–22
additivity of effects, 31, 86–87, 123–24
algebra, cognitive, 210, 214, 223
algebraic models, "as-if" character of, 211
algebraic representation of interaction, 36
aliases, 198–99, 201, 203
alpha level. *See* significance level
ALPHAK computer program, 104, 112
alternative hypothesis, 9
  defined, 247
analysis of covariance, 98n.3, 153
analysis of variance (ANOVA), 3–4. *See also specific topics*
  computation of, 14–15
  example of one-way, 15–16
  model underlying, 4–6
    use of, 6–7
  numerical details of, 16–17
  responsiveness of, 17
  technical assumptions underlying, 123–24
arcsin, 125, 217
"as-if" character of algebraic models, 211
association, 130–31

attrition, 162, 173, 174, 176, 186
averaging model, 222–23, 244n.2

between groups, 9, 10, 14
bilinear analysis, 224, 232–33
blocking, 87–90
Bonferroni procedure, 71, 105

CALSTAT computer programs, 22, 251–52
  installation instructions for, 24
  requirements for and properties of, 24
  exploring the programs, 24–26
cell, empty, 169, 173
cell sizes, proportional, 166–68, 170, 176
change scores, 18, 153
clinical significance, 136–37
coefficients
  for multiplying model, 224, 226
  for specific comparisons, 99–104, 106–8, 110, 111, 197
cognitive algebra, 210, 214, 223
COMPARISONS computer program, 102, 104, 106–8, 110, 112, 113, 116
computer. *See also specific computer programs*

267

computer (*continued*)
  using the, 22–24
confidence interval, 130
  defined, 247
confounded, defined, 30, 247
confounding, 10, 172, 190–91, 198, 199, 201, 203, 204. *See also* Latin squares
conjoint measurement, 244n.3
counterbalancing, 85
covariance, analysis of, 98n.3, 153
critical value, 11
  defined, 247
crossing of factors, 29, 63n.1, 142, 143, 214
CWS, 59, 141n.2

defining contrasts, 198–99, 201
degrees of freedom (*df*), 13–14, 17–18, 85, 90
  defined, 13, 248
  for denominator, 13
  for interaction, 67
  in Latin square, 193, 195
  limitation on specific comparisons, 99, 103, 104
  for numerator, 13
  pooled in nested design, 144
  reducing from error term, 169
delta (Δ). *See* difference between marginal means
dependent *F* tests, 72
dependent variable, 28
  defined, 248
difference between marginal means (Δ), 53–55

ecological validity, 214
empty cell, 169, 173
error term(s)
  for confounded designs, 193–95, 200–203, 207
  for independent groups, 83
  inflated, 85
  for Latin square, 193–95
  for nested designs, 145, 147, 149, 152, 153
  purifying by blocking, 87–90
  reducing degrees of freedom from, 169
  for repeated measures, 83, 85–87
  for specific comparisons, 101, 108, 110–11
error variance, 9
  zero, 124
eta-squared (η^2), 133–34
explanation, 36
explication, 36, 55
extrapolation beyond range of levels used, 36, 109

F ratio, 11
  inappropriately used as strength of effect index, 128
F table, 13, 245
factor, 28
FACTORIAL ANOVA computer program, 32, 36–40, 51, 70
  nested designs and, 144, 153
factorial computer programs, 36–40. *See also* FACTORIAL ANOVA computer program
factorial designs, fractional. *See* fractional factorial designs
factorial plot, 32–35, 53, 215, 233
factorial structure, 28–30, 36
  interaction and, 30–32
  range limitation and, 35–36
factorial symbol, 54
file, 23
fixed effect, 98n.2
fractional factorial designs, 196
  computations for, 199
  error terms in, 200–203
  evaluation of, 203–4
  fractionating a design, 196–98
  multiple defining contrasts in, 198–99
fractional factorials, terminology of, 198
fractionating. *See* fractional factorial designs
functional measurement, 210–11
  general strategy for, 214
    factorial design, 214–16
    scaling, 217
    transformation, 217–18
  models for
    adding model, 218–22
    averaging model, 222–23
    choosing and implementing a model, 229–30

multiplying model,
    223–29
FUNCTIONAL MEASUREMENT
    computer program, 214, 217, 227,
        230–33
    functional measurement diagram,
        211–14
    functional scale, 217
    FUNPOT computer program, 125, 218,
        231, 237

gain scores. *See* change scores
general linear model, 6
generalizability, 58, 83, 98n.2, 109,
    143, 166
grand mean, test of the, 17–18
graphing. *See* factorial plot
Greco-Latin squares, 193–94
group, 8
    defined, 248
group analysis in functional
    measurement, 216
group sizes, unequal. *See* unequal group
    sizes

harmonic mean, 165
Hay's ω^2. *See* omega-squared (ω^2)
high-way designs, 70–71
homogeneity of variance, 123
HSD (Honestly Significant Difference)
    procedure, 114

imputation, 170, 173
independence, 58–59, 120n.1, 124, 216
    randomization and, 11–12
    of scale values, 217
independent groups, 82–83, 132, 135–36
independent-groups design, 161–68
independent variable, 8
    defined, 248
interaction(s), 6, 30–32
    algebraic representation of, 36
    anatomy of, 51–53
    computation of, 49
    in functional measurement, 215, 224
    interpretation of, 68–70
    as possible scale artifact, 56
interval scale, defined, 248
invalid, defined, 248

Latin squares, 191–96
    dangers and inconveniences in, 194
levels, 28
linear fan, 224, 226, 228
linear model, 6
logarithmic transformation, 126–27

main effect, 33, 36
    computation of, 48
    in functional measurement, 215
    interpretation clouded by interaction,
        55–56. *See also* interaction(s)
marginal means scaling, 217–19
mean, test for. *See* grand mean
mean squares (MS), 15
missing scores, 161–63
mixed design, 142
MONANOVA computer program, 125,
    218, 231
monotone transformation, 125, 127, 213
multifactor designs, 64–67, 71–72
    high-way designs, 70–71
    interpretation of interactions in, 68–70
    randomization reemphasized in, 72–73
    simple effects in, 72
multiple comparisons, 104–5, 120n.2
multiplicative model. *See* multiplying
    model
multiplying model, 212, 223–29

naming sources, 37, 47. *See also*
    aliases
nested designs, 142–52
    change scores and, 153
    simple effects in, 153
    used in functional measurement, 216
nesting, 63n.1
    defined, 142
nominal scale, defined, 248
nonorthogonal analysis, 47
nonorthogonal ANOVA, 188n.1
nonorthogonal comparisons, 120n.1
normality, 123
null hypothesis, defined, 9, 248
numerical accuracy, 16–17

omega-squared (ω^2), 130–31
    evaluation of, 131–34
    partial, 132–33

one-tailed test, 11
  defined, 248
ONEWAY ANOVA computer program, 24–27, 56
ordinal scale, defined, 248
ORPOCO computer program, 108
orthogonal polynomial coefficients, 106
orthogonality, 106, 108, 114, 117, 170
  property of, 47
  test, 103–4

p value, 11
  defined, 249
parallelism, 32–33
partial analysis, 71
  evaluation of, 71–72
partial omega-squared (ω^2), 132–33
personal equation, 5
planned comparisons. See specific comparisons
plot. See factorial plot
POLYCON computer program, 125
pooling of sources, 145
post-hoc test, 112
power, 13–14, 84–85, 105, 130
  defined, 13, 249
  in functional measurement, 215
  post-hoc, 137
  of specific comparisons, 99
power transformation, 125
presentation of results, 15–17, 55
pretest-posttest design, 153
probability distribution, 11
  defined, 249
proportion of variance, 132–33, 215
proportional cell sizes, 166–68, 170, 176
psychological integration, 211, 212
psychomotor law, 211, 212
psychophysical law, 211, 212

random attrition, 162, 173
random factor, 98n.2
RANDOM PERMUTATION computer program, 13, 27
randomization, 12–13, 72–73
  in Latin square design, 195
randomized blocks, 87–89
randomness assumption, test of, 162–63

range effects in repeated-measures designs, 98n.1
range limitation, 35–36
ratio scale, defined, 249
repeated measures, 82–87
  additional experimental factors in, 85–86
  assumed additivity in, 86–87, 123–24
  recognition of, 87
  specific comparisons in, 110–11
repeated-measures designs
  accidentally missing scores in, 169–70
  comparisons in, 110–11
  systematically missing scores in
    snapshot analysis, 176–78
    unequal group sizes, 170–73
    zero implantation, 173–76
replacement statement, 145, 147
replicates, 37–38
rescaling, 217, 233
robustness, 123, 124, 188n.1
rounding, 16

scale convergence, 127, 213
Scheffé test, 112–14
SCHUBRAD computer program, 135
Schumann and Bradley test, 134–36
  evaluation of, 136
screen data, 126
significance, clinical, 136–37
significance level, 3, 13, 55
  defined, 249
  set via Bonferroni procedure, 105
significant difference, 55, 114
  defined, 249
significant F ratio, 11
simple effects, 56–58, 86
  in multifactor designs, 72
  in nested designs, 170–73, 176, 177
single-subject design, 58–59, 75, 80, 127–28
small n experiments, 128
snapshot analysis, 176–78
SNAPSHOT computer program, 177
source, 14–15
specific comparisons, 99
split-plot design, 142–43
splitting contrast, 198–201, 205, 206

square root transformation, 125
subject variable, 12
subjects, 8
subscript notation, 4
  defined, 249
substantive source, 10
substantive variable, 10
  defined, 249
subtracting model, 219
sum of squares (SS), 15, 16, 18
  as effects, 53–55
summation, 36
  defined, 250

template for ANOVA computations, 49
tests, too many, 71
transformation, 124–27
  in functional measurement, 217–18, 231
trend coefficients, computer-generated, 108
trend tests, 105–8
  limitations of, 109
Tukey test, 114
two-tailed test, defined, 250
two-way ANOVA, 47–51
  anatomy of interaction in, 51–53
  cautions regarding, 55–56
  presenting the results of, 55
  simple effects in, 56–58
  sums of squares as effects in, 53–55

Type I error, defined, 13, 250
Type II error, 13
  defined, 13, 250

unequal group sizes
  in multi-factor design, 47
  in one-way design, 9
  in repeated-measures design, 170–73
unweighted means, 163–65

validity, 122–23, 162
  in functional measurement, 210–14
  of statistical inferences, 99
variable, 28
variance. *See also specific topics*
  defined, 250–51
  homogeneity of, 123
  proportion of, 132–33, 215
variance diagram, 10

weighted means, 166–68
within-between design. *See* mixed design
within cells, 33
within-cells sum of squares, 49
within groups, 14, 33
within-groups variance, 9

zero error variance, 124
zero implantation, 173–76